The Foundations of Physical Law

Peter Rowlands
University of Liverpool, UK

The Foundations of Physical Law

World Scientific

NEW JERSEY • LONDON • SINGAPORE • BEIJING • SHANGHAI • HONG KONG • TAIPEI • CHENNAI

Published by

World Scientific Publishing Co. Pte. Ltd.

5 Toh Tuck Link, Singapore 596224

USA office: 27 Warren Street, Suite 401-402, Hackensack, NJ 07601

UK office: 57 Shelton Street, Covent Garden, London WC2H 9HE

Library of Congress Cataloging-in-Publication Data

Rowlands, Peter, author.

Foundations of physical law / Peter Rowlands, University of Liverpool.

pages cm

Includes bibliographical references and index.

ISBN 978-9814618373 (hardcover : alk. paper)

1. Physics. 2. Mathematical physics. 3. Quantum theory. I. Title.

QC21.3.R68 2015

530--dc23

2014018315

British Library Cataloguing-in-Publication Data

A catalogue record for this book is available from the British Library.

In-house Editor: Ng Kah Fee

Typeset by Stallion Press

Email: enquiries@stallionpress.com

Preface

The success of physics as the ultimate explanation of everything in the world around us has been phenomenal, but we still have a major problem: we can't formulate the story from the beginning. We have laws of physics, fundamental particles, and concepts like space and time, but we don't have any understanding of their foundations. We don't know where they come from or why they are there. This, of course, is not for want of trying: the search for the ultimate foundations of human knowledge of the world is one of the biggest quests ever undertaken, and it is certainly one of the most expensive, requiring massive international collaborations and incredibly sophisticated facilities. On the experimental side, the search has been successful beyond all expectation, but the results have been disappointing to many because they have repeatedly confirmed existing theories — quantum mechanics, the Standard Model of particle physics, general relativity — without suggesting any new direction leading towards a more general unification. We seem to be no nearer to solving the foundational problems than we were when the Standard Model of particle physics first reached its most complete form around 1973.

I am not convinced that finding 'new physics' would actually help to solve the problems of the old, and I don't think our failure to date has anything to do with the lack of resources or with the subject's inherent difficulty. I think it is a question of mind-set. It seems to me that we may have been asking the wrong questions or perhaps asking the right questions in the wrong way. We haven't properly accepted that if tried and trusted techniques fail to give us answers, then perhaps we need to try new ones. In that sense, the primary aim of this book is to devise a methodology for a subject that we would like to exist but have not yet really attempted to create.

It is remarkable that, for all its importance, the foundations of physics has virtually no recognition as a subject in its own right. It has never been seen as a branch of physics of equal importance as quantum mechanics, particle physics or relativity. It has no place in most university physics courses, and has no legion of PhD students working out details to be tested by experiment. For whatever reason, the subject is not entirely 'respectable'. You can't do it in the hard-edged way that most of physics is done, and so it is very difficult to write a paper that the most prestigious journals would accept. Even though John Bell and others have made it acceptable to talk in a certain way about the foundations of quantum mechanics, it still doesn't take us very far in explaining the foundations of physics itself. Quantum mechanics is still a given in all the discussions, it isn't explained in a more fundamental way. In fact, one of the problems with our failure to define the foundations of physics as a separate subject is the fact that many people think that foundational results can be obtained by extending existing areas, such as quantum mechanics, particle physics, general relativity or cosmology. Experience tells us that they cannot. Foundational ideas have to be simpler than the subjects they are explaining. Extending existing ideas that are already complicated will not bring us nearer to the more primitive starting points.

Nevertheless, for all the general lack of appreciation of the kind of theory needed to explain the foundations of physics and physical law, the desire for such a theory is very strong. My own experience is that students usually enter physics because they really want to know where everything comes from, and that a course which really explains how the search for the foundations should be carried out would come the closest to fulfilling this need. And it is not only physicists and physics students who would benefit from such knowledge for the desire to understand origins and fundamental explanations is part of a much more general human requirement. Clearly, a book or a course of lectures on such a subject, if it reached a deeper level than current investigations would seem to allow, would be of interest to a great number of people.

So, what can one say that is new and positively useful if so many attempts have failed to make any headway? Perhaps we could begin by defining a few negatives — things we want to avoid. The list would certainly include, among other things, major challenges to existing theories, model-dependent ideas, theories that are more complicated than the ideas we are trying to explain, and ideas which are not sufficiently generic.

We could also focus on some ideas which seem to reflect successful previous approaches: minimalism, a high (and preferably total) degree of abstraction, simplicity, symmetry and recurring patterns. In effect, our experience suggests that we need to think in a way that maximises the second list and minimises the first. Even if we adopt such a programme, there is no guarantee that we will be successful, but I would like to suggest that there isn't even a possibility of success if we don't. Nevertheless, this is still not fully understood. Theories are still being produced which do not respect the principles of minimalism and abstraction. By consciously setting out on a path which privileges them, we are already formulating a methodology which, even in this very generalised and simplified form, is diverging from many of the existing approaches towards fundamental theory.

The aim of this book is to show that such a methodology can lead to successful results. Even in purely conventional terms, the version of quantum mechanics that emerges from more abstract foundations using arguments based largely on symmetry appears to be more powerful and effective than alternative versions. Many significant results in physics follow purely from relatively simple arguments privileging symmetry. A generic kind of physics can be constructed from a combination of pure abstraction, symmetry and mathematics without any model-dependent or empirical input. Even gravity and particle physics can be treated in this purely abstract way.

It seems to me important that we take the most successful generic theories and try to find their constituent elements, rather than treat them as elementary and fundamental in themselves. Ultimately, there should be no separate 'physical' constituency to which mathematics is merely applied. At the fundamental level, physics has to be about the simplest ideas, and, almost by definition, these must be the most abstract. This necessarily forces us in the direction of pure mathematics, although (as will be proposed at the end of the last chapter) this mathematics may incorporate ideas which have yet to be written down.

We should also expect even familiar areas of physics to look different in a fundamental context while preserving their standard forms at the more complex levels at which they are currently known. This is especially true of general relativity, which, as Einstein developed it, became a purely abstract theory not based on any specifically *physical* assumptions, and the structures of fundamental particles where the empirically-observed properties clearly have an abstract generic foundation in symmetry principles which have not yet been completely uncovered. As in other areas, the proposed treatment of these subjects at the fundamental level leads to extremely interesting possibilities at the level of observation.

In the case of general relativity, the abstract treatment in Chapter 8 leads to the conclusion that the validity of the theory will extend well beyond the limits currently assumed, together with a prediction of dark energy produced long before its experimental discovery which is very close to the currently observed numerical value. The suggested abstract symmetry structure for fundamental particles in Chapter 9, generated by the convergence of four quite different approaches, makes predictions concerning grand unification within the range of future experimental verification using large colliders.

While the work described in these chapters forms part of a long-term research project whose most significant publication is a book entitled *Zero to Infinity*: The Foundations of Physics, published by World Scientific in 2007, the immediate origin was a course of lectures given in the Physics Department at the University of Liverpool, which were attended by undergraduate and postgraduate students in addition to members of staff, and which were an experiment towards seeing if a course on what was essentially an entirely new subject would generate the expected interest. The book is intended at creating a convenient basis for those who wish to work in the subject or to develop their interests in this area.

Before 1997, the overall research project was a solitary enterprise and much of it still is. This was the origin of the material outlined in Chapters 1–4, and the basis of the work on quantum mechanics, particle physics and gravity in Chapters 5–9. After that period, John Cullerne collaborated with me on aspects of quantum mechanics and particle physics. At the same time, I began to work with Bernard Diaz on the universal rewrite system, and with Peter Marcer on a computational systems approach, both of which appear in Chapter 10. A further extensive collaboration with Vanessa Hill on biology is largely discussed elsewhere, but the table on pp. 202–3, which here refers to fundamental particles, is a result of our discussion of a generic algebraic-geometric patterning observable in many aspects of nature.

The details of these, and several other collaborations, are discussed in *Zero to Infinity*, but I would like to draw attention to the enrichment of the work that they have produced. I would also like to mention the continuing support from a number of members of Liverpool's Physics Department, in particular from Mike Houlden and John Dainton. I would finally like to thank Dave Law and Christian Faber, who filmed and edited the lectures delivered and have made them available to anyone who is interested.

Peter Rowlands
Physics Department
University of Liverpool
June 2014

Contents

Chapter 1

Introduction to Foundational Physics

1.1 What do we mean by foundations of physics?

Investigating the foundations of physics is a risky business. This is a subject without status, career structure, financial support, impact, prizes, journals (despite the existence of one naming itself *Foundations of Physics*), and even protocols. For many people in physics, the subject simply doesn't, or maybe even *shouldn't*, exist. So why study it? The answer is simple: it's the cutting edge. It really is the frontier. That is why systems are not in place, why it has yet to find a place as an integral and essential part of physics. Because physics is a subject in which we are continually trying to push back to the 'origin' from a more complicated position, we can't lay down a generally-accepted basis until we reach the final discovery. When the final goal is to find the starting point, you have a problem setting up a subject that conforms to the usual canons of scientific method, which is mostly structured in precisely the opposite direction.

Despite this apparent problem, we have to persist. Essentially, the most comprehensive model we have to date: the Standard Model of particle physics, has been brilliantly successful. But, as everyone recognises, it is only a *model*, not a theory. It gathers all the facts into a coherent structure, but without any explanation. And the Standard Model has been in place since 1973. We have had no major theoretical advance for over forty years. What is the problem?

In my view, it is because we have confused the search for a *foundational* theory with the search for a *unified* theory. We have put our money on finding a combination of quantum mechanics with general relativity in a large overarching model like the Standard Model. We have imagined that some great superstructure, such as string or membrane theory, will somehow resolve everything, perhaps triggered by an experimental discovery that requires 'new physics'. But, even if we make new experimental discoveries, how will creating a complicated superstructure explain why we have such concepts as space and time to begin with? How will a 10-dimensional theory explain why we have dimensions at all? Even at a more basic level, how does the combination of space and time in both special and general relativity explain why space and time have fundamentally different properties? Combination theories are not asking the right questions. If we want to truly reconcile general relativity and quantum theory, we have to find the foundations on which they are structured.

Foundations of physics is not particle physics, nor general relativity, nor quantum mechanics, nor any kind of extension or combination of them. It is a search for the explanations of such theories and the things that explain them. It is a search for the common origins of all physical theories, and their common origin with mathematics, in effect the explanation of the 'unreasonable effectiveness' of mathematics in physics and of physics in mathematics. If we make progress in this direction, then we will certainly be able to explain many things in classical physics, relativity, quantum mechanics and particle physics, and possibly areas of science of greater complexity. But this is like finding the technological consequences of blue skies research. We know that it will always happen, and we are happy to see it happen, but it is not the research's primary purpose.

1.2 How do we study it?

Foundations of physics is not only a separate discipline within physics, it also requires a completely separate way of thinking and methodology, one that is also intrinsically difficult and has yet to be included in any formal description of the scientific method. In principle, the method exists in that scientists have always used *induction* to infer a cause given certain consequences, but, in studying the foundations, this inductive approach needs to be taken to an extreme level. We have to find causes that are much more general than the ones we usually investigate, and this requires a much more imaginative and wide-ranging style of thinking than our training in the method of deduction from accepted starting points, with perhaps a few carefully-limited inductive inferences, will allow. It means 'thinking outside the box' where the 'box' is physics as we have been taught to understand it. It *doesn't* mean trying to contradict this physics — a generally futile exercise — rather, it means trying to find something which may *look* different, and will certainly be simpler, but which automatically will give us the familiar structure when we make the connections. If we don't recognise this, then we have no hope of making progress.

Even if we do recognise it, there is no obvious way of going about implementing the process. Random speculation isn't likely to be successful if we don't have protocols. In effect, we have to have a theory of knowledge which is more fundamental than the knowledge we want to generate. We have to have a *philosophy*, a 'meta'-physics (in the sense of a more fundamental theory which explains how physics operates), perhaps even a 'metaphysics' in the usual sense. Now, it is quite common to hear physicists talk about 'philosophy' as the kind of thing you do when physics doesn't help you any longer — for example, how do we explain quantum mechanics? is there a First Cause? etc. There is also a whole area of discussion about subjects such as time, the clock paradox, quantum correlation, which is described as the 'philosophy of physics', and another area of discussion about how scientists actually operate called the 'philosophy of science'. I am not talking about any of this when I say 'philosophy'. What I mean is strictly the 'philosophy of knowledge', which is precisely geared to working in the foundations of physics. It is a highly technical process which aims to find those more fundamental principles of knowledge which will help us to choose fundamental principles for physics, and recognise them when we see them.

Now, no one would presume to state a set of general principles of knowledge and then hope that they would work for physics as we know it. The principles would have to emerge from some symbiotic process in which they were developed along with the physics in a continual feedback loop. If, however, the process eventually produced results that we recognised as being correct or significant, it would *then* be possible to begin to codify the principles for use in other cases. This is precisely the spirit in which I intend to propose to outline a methodology or 'philosophy' for investigating the foundations of physics, based on many years of working in this area.

The first thing is to try to get some idea of how we have actually already managed to develop a picture of the world which is very different from the one we normally perceive, and often contradictory to our expectations. To develop a capability of understanding nature, we have had to evolve as highly complex beings in an equally complex environment. We perceive the world at a level which is about 15 orders of magnitude bigger than the size of the proton and neutron, the main constituents of ordinary matter. Though even these are themselves structurally complex, the complexity they create in helping to produce the world as we perceive it is on a stupendous scale. We have no hope of ever reproducing it with any kind of exactitude. Complexity leads to *emergent* properties not seen at a less complex level, and these properties are the ones that we will naturally consider as 'normal', but hundreds of years of investigation have shown that we can be deceived. For centuries, one of the most certain ideas we had was that there was such a thing as solid matter and substance, but now we know that solidity and materiality are only emergent properties. Extended objects don't exist at the fundamental level; 'material' objects appear to be just a distribution of interacting points in an otherwise empty space.

The question is how did we manage to develop this view when our own experience of the world is so completely different? And the answer seems to be that we have been able to use the one real talent that we have — the one without which we couldn't have evolved as a successful species in the struggle for existence. Individual survival and procreation would have been impossible without *pattern recognition*, and the reason why it has been so useful to us is that nature seems to reuse the same patterns at different levels of complexity, a self-similarity and perhaps a fractal structure which is built into its very code. Because of this, we have been able to use patterns observed at one level to explain what occurs at another, as well as at the same time observing more sharply those aspects in which the pattern does not recur. Mathematics is a classic example of self-similarity, a kind of coding of the whole process. But the process has also been crucial in the development of the direct description of the real world in physics, the inherent self-similarity in physics being one of the reasons why we can use the mathematical coding.

Clearly, in looking at the foundations, we need to see pattern as somehow a very significant component — and my own experience is that a highly-developed sense of pattern recognition is one of the most important components in 'thinking outside the box'. We also know from centuries of collective experience that mathematical structures must lie at the heart of physics, but that they must somehow reflect the transition from relative simplicity at the foundational level to more highly developed ideas as more complex structures emerge from the foundations. It is no use producing a highly complex mathematical model as a *foundation*. It might lead to a few successful results in calculation, but, like Ptolemy's epicycles, it can't be the 'real' answer. On the other hand, whatever 'simpler' mathematical structures we use must be capable of leading by a natural progression to the more complex ones already in place. Somehow the mathematics and physics must be intimately connected — mathematics can't simply be a 'tool' used in physics. It has to be deeply built into the structures that physics needs for its foundations.

Simplicity is also key. The progression is always from simplicity at lower levels to complexity at higher levels, never the other way. Explaining complex systems using complex mathematics is an important art, and we have to be able to use it when the foundations are in place, but it doesn't help us to understand the foundations themselves. However, in looking for simplicity, we have a very significant clue which has emerged very strongly in the last few decades. One particularly important type of pattern has become of immense significance, and it has to be an indication of how our mathematics can start from something seemingly simple and lead to something clearly complex. This is *symmetry*, which we see all about us in the laws of physics, the fundamental interactions and the fundamental particles. Finding symmetries can help us to decomplexify our explanations, and, if we can understand where symmetry comes from, lead to more profound understanding. We should also note that some symmetries are *broken*; that is, what is fundamentally symmetric appears, under certain conditions, to display some asymmetry. The reason for this cannot be arbitrary, and if we can discover it, along with the reason for symmetry, this will be a big step in our understanding of the foundations. One thing it can't be is *fundamental* because nature never acts in such an arbitrary way. It has to be, in some way, a sign of complexity or emergence.

1.3 Avoiding the arbitrary

In fact, avoiding the arbitrary is one of the cardinal principles of this way of thinking. Ultimately, you can't subscribe to physics only up to a point. It has to be the total and final explanation. There is no point at which you say 'this is where physics ends'. Of course, some people are on record as saying that there will never be a fundamental theory, and others as saying that physics might be different in different 'universes' — whatever they may be. I don't subscribe to any of this at all. There is no point in physics at all, in my view, if we don't believe that it has to be true without exception. Believing in its unqualified truth has always been the source of its unique strength as an explanation of the universe. Whether we can ever find that truth is another question — though I am totally certain that we can.

Physics has been successful precisely because it does not compromise and this should suggest directions in which we should be looking for the foundations. If we decide a direction is the right one we have to be extreme in implementing it, utterly ruthless in pushing it to its limits. It is becoming increasingly clear, for example, that physics, in its foundations, must be totally abstract, exactly as quantum mechanics would suggest. We shouldn't, therefore, be spending time apologizing for quantum mechanics removing the idea of fixed material objects of a finite size 'interacting' with each other (whatever that may mean) from fixed positions in space. The idea that 'real' or 'tangible' 'objects' and purely abstract concepts like space and time can exist simultaneously in our physical picture is a logical impossibility — like having gods and humans in the same story. A foundational picture cannot be based on the simultaneous existence of things that are incommensurable. They have to be all of the same type, and all the evidence, including our experimental results on the point-like nature of fundamental particles, suggests that the 'true reality' is the abstraction, not the 'tangibility', which is only an emergent property at a higher level of complexity. Only complete abstraction makes logical sense, and only complete abstraction can bring about the desired link with mathematics. Quantum mechanics should, therefore, not be regarded merely as a 'calculating device', but as an exact indication of the way physics should be at the fundamental level. This is not only what physics is like, but what it *should be* like. My own investigations have indicated that a fuller understanding of the mathematical formalisms available to us for representing quantum mechanics provides a totally different understanding of what it is really about, and we will discuss this in the later chapters.

Again, if we decide that simplicity is to be preferred over complexity, we should be looking for something that is staggeringly simple, yet somehow capable of generating complexity. If we think our *basic* idea is a complicated construction, say a 10-dimensional space-time, then we have no way of knowing how this breaks up into the simpler component parts that must exist because we have no *fundamental* mechanism for doing this. We should certainly take notice of what the string theorists say about the symmetries required by nature, and we should expect to find them, even those expressed in 10 dimensions, but we should expect to find them by working out how such a complex idea emerges from simpler ones, in which the structures of the *components* reveal them as diverse in origin, the 'brokenness' of the larger symmetry coming from its inherent complexity, not by some arbitrarily-imposed concept of 'symmetry-breaking'. Broken symmetries are a sure signature of complexity, not of simplicity.

Once we start working purely with abstractions, *models* become meaningless. They are needed when we reach complexity because the complex structures are often difficult to understand without simplification and approximation, but no model lies at the foundations of physics, only pure abstraction. So, when we seek to reach this level, we automatically rule out model-dependent theories. At all times, we go for absolute minimalism, and abstraction is, in effect, minimalism in its final state. The famous rule known as Ockham's razor — we go for the idea which makes the fewest assumptions — is best exemplified by privileging ideas in which everything is there only for an abstract reason, with all arbitrary additions, made for our complexity-driven mode of comprehension (what I call the 'story-book' picture), removed.

1.4 Totality zero

There is one more extraordinary twist to the story. We have removed everything except physics. We have removed everything from physics except abstraction. Surely, we can now start from the abstractions. However, before we can do this, we must apply the principle of minimalism to the *abstract ideas themselves.* We ask the question: is nature at the foundational level characterised by them? And the only possible answer has to be no! Even the abstract concepts, simple as they are, are too arbitrary to be accepted as fundamental principles. The only possible statement that we can make is that nature cannot be characterised — it has no defining characteristics. Any characterisation will necessarily be arbitrary. Here, now, we really need to think about philosophy. Is there a principle which can encapsulate the position we have arrived at, a metaphysical principle if you like? Fortunately, there is and it's been staring at us in the face in physics for quite a long time: *zero totality*. The universe and everything within it, including all the conceptualising that we can do about it, is absolutely nothing. This is the only position that we can take that makes logical sense. Complete zero is the only concept we can imagine that would not be arbitrary if unexplained.

The idea of zero totality is in no way contradictory to the way physics has been moving in the last fifty years. As long ago as the 1960s, Richard Feynman was commenting on the seemingly remarkable fact that the negative gravitational energy of the Hubble universe effectively cancelled out the positive mass energy. He says: "If now we compare the total gravitational energy $E_g = GM_{tot}^2/R$ to the total rest energy of the universe, $E_{rest} = M_{tot}c^2$, lo and behold, we get the amazing result that $GM_{tot}^2/R = M_{tot}c^2$, so that the total energy of the universe is zero. (It is exciting to think that it costs nothing to create a new particle, since we can create it at the center of the universe where it will have a negative gravitational energy equal to $M_{tot}c^2$.) Why this should be so is one of the great mysteries — and therefore one of the important questions of physics. After all, what would be the use of studying physics if the mysteries were not the most important things to investigate."[1] Cosmologists, similarly, have long spoken about everything starting from 'nothing', or zero space, time and matter. The well-known chemist and science writer Peter Atkins has said that 'the seemingly something is elegantly reorganised nothing, and... the net content of the universe is ... nothing'.[2] Even classical physics is strongly based on the idea that the total force in the universe must always be zero — to every action there is an equal and opposite reaction.

How this would work out at a higher level of complexity would depend on how it was introduced at the foundational level, but at that level there is an obvious way in which it could happen; and this is something we have already discussed. If we are presented with a need to make everything total exactly zero, then we have a ready-made explanation for why symmetry is so important at the deepest level. Symmetry (and, in particular, that form of symmetry known as *duality*) provides a clear way of having totality zero and yet allowing things to happen. This is routine for physicists, who, for example, will say that two objects will fly apart with zero total momentum, and yet each object has a nonzero momentum of its own. Now, if we want to introduce it as a *general* principle, we will have to imagine that the real zeroing happens at a subtle foundational level, and that apparent asymmetry or symmetry-breaking will come with complexity. This can be justified at a later date (and will be) when we look at how complex systems develop from simpler abstract notions, but, for the moment, we will simply accept that this is a principle which, like all the others, applies, at the foundational level, *without exception*.

1.5 What questions should we ask?

We now have in place our methodology. We require our foundational ideas to be simple, symmetrical, mathematically-based, minimal, totally abstract, and combining to a zero totality, and we want these conditions to be applied rigorously, ruthlessly, and in all cases without exception. We want to ensure that nothing enters into the picture which is in any way arbitrary, and that the final 'structure' (for want of a better word) is also *totally exclusive*, that is, that nothing in physics can exist outside of it. Anything breaking these conditions would not be truly foundational.

We have now established strict protocols for a foundational approach, all of which (even the 'metaphysical' principle of zero totality) are perfectly reasonable within the context of physics as we know it. This is only the first step — we haven't yet established what they will apply to — but it is a very important one. At this stage we want to exclude everything extraneous, to reduce the options to the only ones that are compatible with a foundational theory as we perceive it. The strictness is absolutely essential, as we are reaching the absolute limit of what it is conceivable to know.

Now, we need to establish what concepts fulfil these conditions. These will be the most primitive physical concepts imaginable. What we can do here is appeal to our original idea that the whole human development of science is only possible because we have the capacity to recognise repeated patterns. What we seem to find is that, as our investigation of nature descends from the large-scale complex structures to increasingly small-scale and less complex ones, certain concepts seem to be necessary at all levels (for example, space and time), while others (solidity, materiality) are 'emergent' and disappear at the less complex levels. This suggests that the former may appear at the most primitive level, while the latter will not. Using this kind of analysis we can put forward a provisional set of 'primitive' concepts (ones which cannot be broken down any further or discarded as emergent) which we can test against our protocols. The full account will be given in the third chapter after we have established some necessary mathematics.

However, it is clear that they must include space and time, but *not* space-time, which is emergent (otherwise we could not explain the differences between the components — another broken symmetry!), and certainly not curved space-time, which is even more emergent. Clearly, something must represent matter in its point-like state and something else energy or the connections between the points. There must also be some way in which, at the foundational level, these abstract concepts are delivered, all at once, as a 'package', explaining the growth of complexity in the way the packaging occurs.

Whatever we find won't immediately look like physics as we now know it, any more than a cluster of cells looks like a human being, but, as we unravel the complexities, we will see that familiar physics will begin to appear, and with a greater clarity than it ever previously had. People have long had an expectation that some great complex equation will hold all of nature's secrets, but nature doesn't work like that at all. This primitive 'structure' will certainly be mathematical, but it won't be an equation. Its secrets won't be immediately obvious. They will have to be teased out one by one, as the complexity develops. This is physics at a primitive, embryonic level, an *ur-Theorie* if you like. It requires inductive thinking, a kind of X-ray vision which penetrates through the layer upon layer of complexity to the primitive core — deductive mathematical techniques won't help us. But the severity of the criteria we have established and the exact mathematical basis will ensure that it is rigorous. And it will begin to answer those questions, like, "What are space, time and matter?", "Why is space as we observe it 3-dimensional?" and "Why does time never run backwards?", that we may have thought could never be answered.

Chapter 2

Mathematical Ideas and Methods

2.1 Quaternions and octonions

There is a mathematics, vital to us, which is spoken of with respect in some quarters, is recognised as the parent of several other branches of mathematics very much in current use, is used every day by software engineers to represent motion in space and time, and appears in a disguised form in the treatment of relativity and quantum mechanics, yet is regarded with contempt by many who don't know anything about it. Those who don't know their history are condemned to repeat it, and the people who repeat what they have learned from their tutors about the intrinsic worthlessness of this mathematics usually have no idea that they are contributing to the perpetuation of what amounts to an episode of 'cultural genocide' that began more than a hundred years ago.

For physics, the prevalence of this attitude has had disastrous consequences because it has kept us away from the branch of mathematics that is most necessary for understanding its foundations, the only one that has ever given us a handle on the meaning of 3-dimensional space. It is, of course, absurd to claim that any branch of pure mathematics is *intrinsically* worthless and that it will never find valid applications. How can anyone know? In any case, the current massive use in the commercially-driven software industry disproves the claim many thousands of times over. Yet, it has meant that this branch of mathematics is taught very little in physics courses, making the intrinsic lack of utility a self-fulfilling prophecy. This is, quite frankly, the biggest missed opportunity in the history of the subject.

It was back in 1843 when Sir William Rowan Hamilton realised, for the first time, that he had discovered the meaning of 3-dimensionality. He was trying to extend the idea of complex numbers to the next level. He reasoned that this could be done by adding a third axis to the well-known Argand diagram. On the Argand diagram, the real numbers, based on unit 1, are on the horizontal or x-axis, the imaginary numbers, based on unit $i = \sqrt{-1}$, are on the vertical or y-axis. Hamilton proposed drawing a third, z-axis perpendicular to the other two, which couldn't contain real numbers, as these are all on the x-axis, but could conceivably contain another set of imaginary numbers based on a unit $j = \sqrt{-1}$ which was different from i. The only problem with this is that it doesn't work algebraically. We can find products of numbers with units 1 and 1, 1 and i, 1 and j, i and i, and j and j, but we can't find a product of numbers with units i and j. It can't have real units, because this would equate i and j; it can't have unit i, because this would equate 1 and j; and it can't have unit j because this would equate 1 and i. The system doesn't exhibit *closure*.

After many years of struggle, he found a solution, but this meant violating one of the cardinal principles of algebra as then known: the principle of commutativity. In effect this meant that if you took the product *ba*, that would produce exactly the same answer as taking the product *ab*. What Hamilton did was to remove the real axis entirely and have 3 imaginary axes, with units $\boldsymbol{i}$, $\boldsymbol{j}$, $\boldsymbol{k}$, all different from each other but all equating to $\sqrt{-1}$, and following a rotation cycle, so that

$$\begin{aligned}\boldsymbol{i}^2 = \boldsymbol{j}^2 = \boldsymbol{k}^2 = \boldsymbol{ijk} &= -1\\ \boldsymbol{ij} &= \boldsymbol{k}\\ \boldsymbol{ki} &= \boldsymbol{j}\\ \boldsymbol{jk} &= \boldsymbol{i}\end{aligned}$$

This works well but it requires a reversal of sign when we reverse the order of multiplication. That is:

$$\begin{aligned}\boldsymbol{ji} &= -\boldsymbol{k}\\ \boldsymbol{ik} &= -\boldsymbol{j}\\ \boldsymbol{kj} &= -\boldsymbol{i}\end{aligned}$$

The units are *anticommutative*. The reason can be seen immediately. Let us, for example, take the product $\boldsymbol{ijji}$. We multiply $\boldsymbol{j}$ with itself first, and their product (or norm) is -1. The remaining term is $-\boldsymbol{ii}$, which clearly equals 1. So

$$\boldsymbol{ijji} = -\boldsymbol{ii} = 1$$

But this can only be true if $\boldsymbol{ji} = -\boldsymbol{k}$. Accepting this had to be done, Hamilton then had a closed algebra, with four basic units 1, $\boldsymbol{i}$, $\boldsymbol{j}$, $\boldsymbol{k}$, which he called *quaternions*, and which was double the size of ordinary complex algebra with units 1, i, and four times the size of real algebra, based on unit 1.

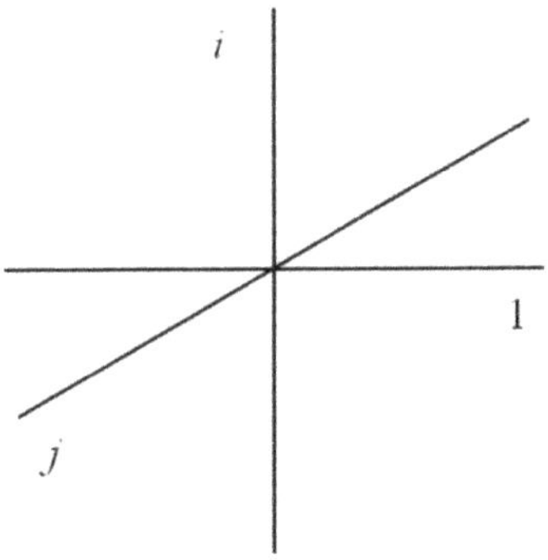

Extended Argand diagram.

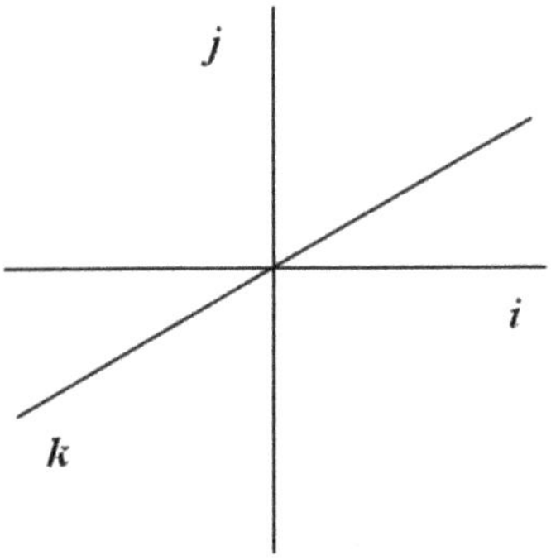

Quaternions.

The next question he asked was could we increase the size of this algebra and still maintain the same rules, say with one real and four imaginary units: 1, $\boldsymbol{i}$, $\boldsymbol{j}$, $\boldsymbol{k}$, $\boldsymbol{l}$? And the answer is we can't. You can't get consistency if you extend the number of imaginary units beyond three. Hamilton suspected this was the case; Frobenius proved it in 1878. There is just one exception. You can create a consistent system with one real and *seven* imaginary units, 1, $\boldsymbol{i}$, $\boldsymbol{j}$, $\boldsymbol{k}$, $\boldsymbol{e}$, $\boldsymbol{f}$, $\boldsymbol{g}$, $\boldsymbol{h}$, called the *octonions*. However, to do this you have to break another algebraic rule, the law of associativity in which, say, $(ab)c$ is always equal to $a(bc)$. So, octonions, unlike quaternions, are antiassociative as well as anticommutative, leading to products equivalent to $(ab)c = -a(bc)$.

The trick can't be repeated at any other level, so we are left with just four so-called division algebras:

Real	norm 1	commutative	associative
Complex	norm –1	commutative	associative
Quaternions	norm –1	anticommutative	associative
Octonions	norm –1	anticommutative	antiassociative

If we are looking for quaternions, as Hamilton did, as the potential explanation for the 3-dimensionality of space, the thing to note is that the '3-ness' isn't the primary cause. It is simply a result of anticommutativity. If we have two axes, $\boldsymbol{i}$ and $\boldsymbol{j}$, that are anticommutative with each other, then we cannot draw any other axis that is anticommutative with them, unless it is $\boldsymbol{ij}$, which we also call $\boldsymbol{k}$. Anticommutativity forces 3-dimensionality. The strange arbitrariness of the number 3 is explained. Commutative things, of course, can be defined to infinity. If $\boldsymbol{i}$ and $\boldsymbol{j}$ were commutative, we could have $\boldsymbol{i}$, $\boldsymbol{j}$, $\boldsymbol{k}$, $\boldsymbol{l}$, $\boldsymbol{m}$, etc. without limit. In a sense we can say that anticommutative things 'know' about each other's presence and have to act accordingly; commutative things do not.

Now, just as a complex number is fully represented by the sum of a real part and an imaginary part, say, $x + iy$, so a quaternion number will have a real and three imaginary parts, say, $\boldsymbol{a} = w + \boldsymbol{i}x + \boldsymbol{j}y + \boldsymbol{k}z$, where w, x, y, z are just positive or negative scalars or real numbers. (For convenience, we will always represent quaternion units by ***bold italic*** symbols.) Let us now suppose we have another quaternion, say $\boldsymbol{a}' = w' + \boldsymbol{i}x' + \boldsymbol{j}y' + \boldsymbol{k}z'$. If we take the product of $\boldsymbol{a}$ and $\boldsymbol{a}'$, multiplying it out term by term, we will obtain:

$$\boldsymbol{aa}' = ww' - (xx' + yy' + zz') + \boldsymbol{i}(w + w' + yz' - zy')$$
$$+\boldsymbol{j}(w + w' + zx' - xz') + \boldsymbol{k}(w + w' + xy' - yx')$$

Hamilton, who first obtained this product, described w as the *scalar* part of the quaternion $w + \boldsymbol{i}x + \boldsymbol{j}y + \boldsymbol{k}z$, and $\boldsymbol{i}x + \boldsymbol{j}y + \boldsymbol{k}z$ as the *vector* part. He also described $ww' - (xx' + yy' + zz')$ as the *scalar product* and $\boldsymbol{i}(w + w' + yz' - zy') + \boldsymbol{j}(w + w' + zx' - xz') + \boldsymbol{k}(w + w' + xy' - yx')$ as the *vector product*. We may note the similarity to the use of these terms in vector theory today. He also introduced a quaternion differential operator, $\nabla = \boldsymbol{i}\partial/\partial x + \boldsymbol{j}\partial/\partial y + \boldsymbol{k}\partial/\partial z$. He speculated that the real or scalar part of the quaternion represented time and the imaginary or vector part space, and that the quaternion structure showed the long-sought link between them. Hamilton also realised that unit quaternions could be used to represent rotations in 3-dimensional space. He was convinced that he had discovered the reason why space had to be 3-dimensional and that quaternions would be the key to unlocking the secrets of the universe.

It all looked very promising, and Maxwell provided alternative quaternion treatments of mathematical operations in his famous *Treatise on Electricity and Magnetism* in 1873. But then it all went wrong. In the late nineteenth century, after both Hamilton and Maxwell were dead, people started complaining about the fact that quaternions, when squared in Pythagoras' theorem, produced the wrong sign of product, negative instead of positive. They also disliked the connection between the real and imaginary parts, preferring a structure that had just three real parts and no imaginary part. Ultimately, Gibbs and Heaviside formulated a new *vector* theory, which made the 3-dimensional or vector part real, discarded the fourth component, and recreated the scalar and vector products as two separate operations. Their vector theory was simply a rule book which has been the everyday tool for physicists ever since, though its arbitrariness has always caused problems for students. After all, vector 'algebra' is not an algebra at all as it has no multiplication and its operations do not exhibit closure. Vectors can have two 'products', neither of which results in another vector. The scalar product produces a scalar and the vector product a new type of quantity, a *pseudovector* (or axial vector), which transforms differently to vectors, while the scalar product of a vector and a pseudovector produces yet another type of quantity, a *pseudoscalar*, which is quite different from a scalar.

Unfortunately, the supporters of the new theory decided there wasn't room for both vector theory and quaternions and an intensive vilification campaign led to the annihilation of the quaternionists. Their mathematics was not only less useful than the new vector theory but, in fact, utterly worthless and totally without application. Hamilton's view that quaternions were the key to the universe was one of the most self-deluding ideas ever entertained by a great mathematician. His career, so promising at first, had ended in total tragedy. When relativity came along soon afterwards and made the connection between space and time that Hamilton had sought in his mathematics, so solving one of the two main objections to the quaternion theory, no one wanted to know. The connection, they said, was nothing to do with quaternions. It is hard to find another defeat so comprehensive in the history of science.

Yet Hamilton had not only been right all along about the space-time connection, so removing the second objection, he had already produced the mathematics that would have removed the first objection regarding the norm as well. This was in his very first development of the original idea. Since quaternions were quite distinct from ordinary complex numbers, why not combine the two and produce complexified quaternions? Our base set is now 1, i, $\boldsymbol{i}$, $\boldsymbol{j}$, $\boldsymbol{k}$, and, multiplying everything out, we will also generate terms like $i\boldsymbol{i}$, $i\boldsymbol{j}$, $i\boldsymbol{k}$. For reasons that will soon become clear, I also write $i\boldsymbol{i} = \mathbf{i}$, $i\boldsymbol{j} = \mathbf{j}$, $i\boldsymbol{k} = \mathbf{k}$. If we take the products of these terms, we can write:

$$(i\boldsymbol{i})^2 = (i\boldsymbol{j})^2 = (i\boldsymbol{k})^2 = -i(i\boldsymbol{i})(i\boldsymbol{j})(i\boldsymbol{k}) = 1$$
$$(i\boldsymbol{i})(i\boldsymbol{j}) = i(i\boldsymbol{k})$$
$$(i\boldsymbol{k})(i\boldsymbol{i}) = i(i\boldsymbol{j})$$
$$(i\boldsymbol{j})(i\boldsymbol{k}) = i(i\boldsymbol{i})$$

The complexified quaternion units $i\boldsymbol{i} = \mathbf{i}$, $i\boldsymbol{j} = \mathbf{j}$, $i\boldsymbol{k} = \mathbf{k}$ are, of course, anticommutative in exactly the same way as ordinary quaternion units, but we now notice an extra feature, the i term outside the bracket that has appeared on the right-hand side of the equations. This becomes clearer if we write the equations in our alternative, more compactified, notation:

$$\mathbf{i}^2 = \mathbf{j}^2 = \mathbf{k}^2 = -i\mathbf{ijk} = 1$$
$$\mathbf{ij} = i\mathbf{k}$$
$$\mathbf{ki} = i\mathbf{j}$$
$$\mathbf{jk} = i\mathbf{i}$$

The remarkable thing now is that these objects have the properties that we require of vectors. In particular, they square to positive values. But they have something else in addition, an extra property whose meaning didn't emerge until well into the twentieth century. They also incorporate *spin*, that is, the mysterious property introduced by quantum mechanics. We can recognise them as being isomorphic to the Pauli matrices, originally introduced into nonrelativistic quantum mechanics to incorporate the experimentally-discovered concept of spin.

$$\sigma_x = \begin{pmatrix} 0 & 1 \\ 1 & 0 \end{pmatrix} \quad \sigma_y = \begin{pmatrix} 0 & -i \\ i & 0 \end{pmatrix} \quad \sigma_z = \begin{pmatrix} 1 & 0 \\ 0 & -1 \end{pmatrix}$$

Hestenes, later in the twentieth century, termed **i**, **j** and **k** as the units of a *multivariate vector* algebra.[3] In general, multivariate vectors **a** and **b** followed a full multiplication rule, which incorporated both scalar and vector products in the same way as quaternions:

$$\mathbf{ab} = \mathbf{a} \cdot \mathbf{b} + i\mathbf{a} \times \mathbf{b}$$

He showed that if we used the full product $\nabla\nabla\psi$ for a multivariate vector ∇ (basically, Hamilton's own definition of the symbol!) instead of the scalar product $\nabla \cdot \nabla \psi$ for an ordinary vector ∇, we could obtain spin $\frac{1}{2}$ for an electron in a magnetic field from the nonrelativistic *Schrödinger equation*. Though the first explanation of spin $\frac{1}{2}$ came from the relativistic Dirac equation, the effect is nothing to do with relativity. It comes from properties deep within 3-dimensionality. The reason why Dirac first obtained it is because he effectively included these properties in the extra algebra he needed to make his equation linear.

In fact, all physical vectors are really multivariate vectors and not ordinary vectors at all, and, generally, when I use the word 'vector' I will use the multivariate definition unless there is a specific reason to do otherwise. I will also always use **bold** symbols for vectors to distinguish them from the ***bold italics*** used to represent quaternions and the *italics* used for ordinary complex numbers. Unlike ordinary vector algebra, multivariate algebra is a real algebra. It has closure and a genuine product. It also makes sense of such things as pseudovectors and pseudoscalars, which appear arbitrarily in ordinary vector algebra. As we can see from its multiplication rules, multivariate vector algebra generates such terms as $i\mathbf{i}$, $i\mathbf{j}$, $i\mathbf{k}$ (which are not vector units) and also i (which is not an ordinary scalar). Let's take a simple example. Imagine we have a rectangle with sides $\mathbf{a}$ and $\mathbf{b}$. To find the area, we take the product $\mathbf{ab} = \mathbf{a} \cdot \mathbf{b} + i\mathbf{a} \times \mathbf{b}$. Since $\mathbf{a}$ and $\mathbf{b}$ are orthogonal, the first term on the right-hand side disappears, leaving us with an imaginary vector in a direction perpendicular to $\mathbf{a}$ and $\mathbf{b}$. The area is an imaginary or *pseudovector*, say $i\mathbf{A}$. If we then suppose that the rectangle is the base of a solid body with height $\mathbf{c}$ in the direction of this pseudovector, then the volume will be the product $i\mathbf{Ac} = i\mathbf{A} \cdot \mathbf{c} + i\, i\mathbf{A} \times \mathbf{c}$. This time, since the vector and pseudovector are parallel, it is the second term which disappears, leaving the product as an imaginary scalar. So volume (the 'triple product' in fact, as well as in name) is a pseudoscalar.

Because we require the pseudovector and pseudoscalar terms as well as vectors and scalars, vector algebra is a larger algebra than quaternion algebra. In fact, quaternions can be seen as a subalgebra of vectors, composed of the pseudovectors and scalars. Remarkably, pseudovectors, which include such concepts as torque and angular momentum, are identical in principle to quaternions, and we can (with appropriate sign adjustments) switch from quaternion to vector representations and vice versa simply by multiplying the units by i. In addition, the combination of vector and pseudoscalar units, ($\mathbf{i}$, $\mathbf{j}$, $\mathbf{k}$) and i, in the vector algebra brings us to the recognition that vector algebra in this form also incorporates the 4-vector algebra of relativity, which requires the units ($\mathbf{i}$, $\mathbf{j}$, $\mathbf{k}$, i) and is equivalent to the complexified version of a quaternion (with units $\boldsymbol{i}$, $\boldsymbol{j}$, $\boldsymbol{k}$, 1).

It is amazing how powerful quaternions look in this new context, and we are only just beginning! Was Hamilton also right about quaternions unlocking the secrets of the universe? I think he probably was, and I hope we will see why as we proceed through the chapters.

2.2 Clifford algebra

The most remarkable extension of quaternions comes with the algebra invented by Clifford in the 1870s. In fact it was Clifford who first wrote down the expression $\mathbf{ab} = \mathbf{a} \cdot \mathbf{b} + i\mathbf{a} \times \mathbf{b}$. Clifford algebra is one of the most powerful tools ever offered to the physicist, and this is no coincidence, since, in many ways, it seems to be the mathematical code built deep in the structure of physics. It is still massively under-used, though Dirac recognised early on that the algebra he had devised for his equation for relativistic quantum mechanics was, in fact, a Clifford algebra.

Clifford algebra (also called geometrical algebra) unites real, complex numbers, quaternions and vectors into a single system of infinite potential complexity. It is structured on defining a system with m units which are square roots of 1 (norm 1) and n which are square roots of –1 (norm –1), where m and n are integers of any size. We write this as $Cl(m, n)$ or $G(m, n)$. However, this is not necessarily a unique specification for it is often possible to produce the same algebra with a quite different specification of m and n.

One way of building up Clifford algebras is to use commuting sets of quaternions, which may also be complexified. I could for instance build up a Clifford algebra using three sets of quaternion units, each of which commutes with the others, although the units within the quaternion sets anticommute. I could then create a higher Clifford algebra by complexifying it. The basic Clifford algebra is the Clifford algebra of 3-dimensional space, which we have already described in detail under its later name of multivariate vector algebra. In its full specification it has 16 basic units, versions of which can be + or −:

$\mathbf{i}$ $\mathbf{j}$ $\mathbf{k}$	vector		
$i\mathbf{i}$ $i\mathbf{j}$ $i\mathbf{k}$	bivector	pseudovector	quaternion
i	trivector	pseudoscalar	complex
1	scalar		

It has 3 subalgebras: bivector/pseudovector/quaternion, composed of:

*i*i *i*j *i*k	bivector	pseudovector	quaternion
1	scalar		

trivector/pseudoscalar/complex, composed of:

i	trivector	pseudoscalar	complex
1	scalar		

and scalar, with just a single unit:

1

Here, we use the term 'bivector' for the product of two vectors and 'trivector' for the product of three. It is important for us to recognise that we could specify the entire algebra, either by using (i, j, k) or a combination of its 3 subalgebras. This will become very significant in our work.

A particularly interesting algebra emerges if we combine this algebra with an identical algebra of 3-dimensional space to which this is commutative.

i j k	vector		
*i*i *i*j *i*k	bivector	pseudovector	quaternion
i	trivector	pseudoscalar	complex
1	scalar		

When we multiply these two algebras by each other, term by term, we produce an algebra that has 64 basic units, which are + and − versions of:

i	j	k	*i*i	*i*j	*i*k	*i*	1
i	j	k	*i*i	*i*j	*i*k		
ii	ij	ik	*i*ii	*i*ij	*i*ik		
ji	jj	jk	*i*ji	*i*jj	*i*jk		
ki	kj	kk	*i*ki	*i*kj	*i*kk		

Since vectors are complexified quaternions and quaternions are complexified vectors, we obtain an identical algebra if we use complexified double quaternions:

i	*j*	*k*	*ii*	*ij*	*ik*	*i*	1
i	*j*	*k*	*ii*	*ij*	*ik*		
ii	*ij*	*ik*	*iii*	*iij*	*iik*		
ji	*jj*	*jk*	*iji*	*ijj*	*ijk*		
ki	*kj*	*kk*	*iki*	*ikj*	*ikk*		

Yet another variation can be found using a combination of vectors (blue) and quaternions (*red*):

i	j	k	$i\mathrm{i}$	$i\mathrm{j}$	$i\mathrm{k}$	i	1
i	j	k	ii	ij	ik		
$\mathrm{i}i$	$\mathrm{i}j$	$\mathrm{i}k$	$i\mathrm{i}i$	$i\mathrm{i}j$	$i\mathrm{i}k$		
$\mathrm{j}i$	$\mathrm{j}j$	$\mathrm{j}k$	$i\mathrm{j}i$	$i\mathrm{j}j$	$i\mathrm{j}k$		
$\mathrm{k}i$	$\mathrm{k}j$	$\mathrm{k}k$	$i\mathrm{k}i$	$i\mathrm{k}j$	$i\mathrm{k}k$		

One of the remarkable things about Clifford algebra is that the same algebra can have different dimensionalities simultaneously. It depends on how you distribute between m and n. So two square roots of –1 can become a single square root of 1 and vice versa. And this can be done multiples of times. So, you can have a 10-dimensionality in one perspective simultaneously with a 3-dimensionality in another. Ultimately, as Hamilton recognised at that moment of discovery in 1843, 3-dimensionality is special and is due to a single, simple property. Any higher dimensionality can always be resolved into structures based on 3, and even complex numbers can be represented as incomplete quaternion sets.

2.3 Groups

Groups are an important aspect of the mathematics of symmetry, and, since symmetry is going to be a major aspect of our foundational approach, we can expect groups to play a significant part. A finite or infinite number of elements form a group if they contain:

(1) An associative binary operation (e.g. multiplication, addition) between any two elements to produce another
(2) An identity element, so that the binary operation between the identity element and another element produces that element
(3) An inverse to each element, so that the binary combination of any element and its inverse produces the identity element
(4) Closure, so that the binary operation between any two elements always produces an element within the group

As an illustration, we can take the simplest group C_2, which is of order 2 and so has only two elements, which we could, for example, take as 1 and -1. Here the binary operation is multiplication. The identity element is 1. Each element is its own inverse, and we have closure because each binary operation between 1 and −1 only ever produces either 1 or −1. Complex numbers give us a group of order 4 (C_4) made out of the base units 1, −1, i and $-i$. Quaternions (Q) give us a group of order 8, from the base units 1, −1, $\boldsymbol{i}$, $-\boldsymbol{i}$, $\boldsymbol{j}$, $-\boldsymbol{j}$, $\boldsymbol{k}$, $-\boldsymbol{k}$. Multivariate vectors, which complexify the quaternion algebra, have a group of order 16. In all these cases negative as well as positive units are required for closure. Octonions (O), however, are not a group because their multiplication (see the tables at the end of the chapter) is not associative. For any group, finite or infinite, it is possible to produce all the elements of the group using a finite number of elements as *generators*. For a finite group, this may be less than the number of elements in the group. Often, this can be done in many different ways.

C_2 and C_4 are examples of *cyclic* groups. There is a cyclic group at every order, and how they operate can be illustrated by writing down the multiplication table for C_3.

*	I	a	a^2
I	I	a	a^2
a	a	a^2	I
a^2	a^2	I	a

A cyclic group of order n, has elements, I, a, $a^2, \ldots, a^{n-1}$, with a^n reverting to the identity (I). So, if we wanted to express the complex number units as a cyclic group of order 4, then I, a, a^2, a^3 would become 1, i, -1, $-i$, in that order. Now, there is a second group of order 4 which will be very important to us, which is called D_2 or the Klein-4 group. It is called D_2 or dihedral 2 because it is the group of rotations of the rectangle (identity, and rotations along 3 different axes), and it is the only group other than C_4 of order 4. One representation is through a 'double' algebra known as H_4, which is made up of 4 units, made up of two commutative sets of quaternions, 1, i, j, k, and 1, i, j, k. Here, the red and blue units multiply commutatively with each other in the ordinary way, but the red units anticommute with *each other*, as do the blue units. The H_4 algebra units can be constructed as 1, ii, jj, kk, and you can see that the units ii, jj, kk have now become commutative with each other, unlike the units of their parent systems. So $iijj = jjii$, etc. The H_4 algebra units become a group with the multiplication table:

*	1	*ii*	*jj*	*kk*
1	1	*ii*	*jj*	*kk*
ii	*ii*	1	*kk*	*jj*
jj	*jj*	*kk*	1	*ii*
kk	*kk*	*jj*	*ii*	1

It's effectively the same thing as using quaternions but ignoring the negative signs.

Another way of arriving at the same table is by creating elements made up of variations of three 'components', $\pm x$, $\pm y$ and $\pm z$:

A	x	y	z
B	$-x$	$-y$	z
C	x	$-y$	$-z$
D	$-x$	y	$-z$

The binary operation ($*$) doesn't have to be anything to do with multiplication. So we can make up our own rules as long as they are rigidly followed and lead to closure. Let's say we have:

$$x * x = -x * -x = x$$

$$x * -x = -x * x = -x$$

and similarly for y and z. We also demand that for $A = (x, y, z)$ and $B = (-x, -y, z)$, $A * B = (x * -x, y * -y, z * -z) = (-x, -y, z)$. Then we will have a group table:

*	A	B	C	D
A	A	B	C	D
B	B	A	D	C
C	C	D	A	B
D	D	C	B	A

This version of the group will become even more important to us than the one illustrating the H_4 algebra, and I hope to show later that, overall, there is no more important group in the whole of physics.

We can also arrange for a dual version of this group, with assignments such as:

$\mathbf{A}^\dagger$	$-x$	y	z
$\mathbf{B}^\dagger$	x	$-y$	z
$\mathbf{C}^\dagger$	$-x$	$-y$	$-z$
$\mathbf{D}^\dagger$	x	y	$-z$

Another very significant group for us is the group of order 64 coming from by the commutative product of two vector algebras or of a vector algebra with a quaternion algebra. If we write out the complete set of 64 terms for the vector-quaternion combination, we find that, apart from the scalar and pseudoscalar units, something very interesting emerges.

1	i				-1	$-i$			
$i\mathrm{i}$	$i\mathrm{j}$	$i\mathrm{k}$	ik	j	$-i\mathrm{i}$	$-i\mathrm{j}$	$-i\mathrm{k}$	$-ik$	$-j$
$j\mathrm{i}$	$j\mathrm{j}$	$j\mathrm{k}$	ii	k	$-j\mathrm{i}$	$-j\mathrm{j}$	$-j\mathrm{k}$	$-ii$	$-k$
$k\mathrm{i}$	$k\mathrm{j}$	$k\mathrm{k}$	ij	j	$-k\mathrm{i}$	$-k\mathrm{j}$	$-k\mathrm{k}$	$-ij$	$-i$
$i\mathrm{i}i$	$i\mathrm{i}j$	$i\mathrm{i}k$	$i\mathrm{k}$	j	$-i\mathrm{i}i$	$-i\mathrm{i}j$	$-i\mathrm{i}k$	$-i\mathrm{k}$	$-\mathrm{j}$
$i\mathrm{j}i$	$i\mathrm{j}j$	$i\mathrm{j}k$	$i\mathrm{i}$	k	$-i\mathrm{j}i$	$-i\mathrm{j}j$	$-i\mathrm{j}k$	$-i\mathrm{i}$	$-\mathrm{k}$
$i\mathrm{k}i$	$i\mathrm{k}j$	$i\mathrm{k}k$	$i\mathrm{j}$	i	$-i\mathrm{k}i$	$-i\mathrm{k}j$	$-i\mathrm{k}k$	$-i\mathrm{j}$	$-\mathrm{i}$

Apart from the 4 units of ordinary complex algebra, 1, i, -1, $-i$, the other sixty units arrange themselves in 12 groups of 5. Remarkably, *any* of these groups of 5 will generate the *entire group*, and this is in fact the minimum number of generators. Significantly, they all have the same overall structure. We take the 8 base units provided by a 4-vector (i, j, k, i) combined with a quaternion (i, j, k, 1), and break the symmetry of one of the two 3-dimensional structures to create the minimum structure for the generators. From the perfect symmetry of

$$i \qquad \mathrm{i} \quad \mathrm{j} \quad \mathrm{k} \qquad 1 \qquad i \quad j \quad k$$

we rearrange to produce:

$$\begin{array}{ccccc} i & \mathrm{i} & \mathrm{j} & \mathrm{k} & 1 \\ k & & i & & j \end{array}$$

and finally:

$$ik \qquad i\mathrm{i} \quad i\mathrm{j} \quad i\mathrm{k} \qquad j$$

Here, the symmetry of the *red quaternion operators* is broken in that they are attached to different types of object: pseudoscalar i, vectors i, j, k, and scalar 1. The symmetry of the vector operators, however, is retained because each of i, j, k is associated with the same object. We can imagine that, if we wanted to make the minimum package containing two 3-dimensional structures, the symmetry of one would be broken in this way.

Exactly the same result would have followed if we had used double vectors (or the units of two 'spaces'), where the 5 generators would have been

$$i\mathbf{k} \qquad \mathbf{ii} \qquad \mathbf{ij} \qquad \mathbf{ik} \qquad \mathbf{j}$$

In addition to finite groups there are also infinite or Lie groups, with an infinite number of elements and a finite number of generators. Lie groups are often generated from the rotational aspects of finite groups. The unitary groups, $U(n)$, are defined as the groups of $n \times n$ unitary matrices, and so have a complex determinant with norm 1. A complex square matrix, such as the ones which define $U(n)$ groups, is unitary if the product of the matrix U and its conjugate transpose $U^\dagger$ is the identity matrix I. $U(n)$ groups have n^2 generators. The simplest of these, $U(1)$, can be illustrated by drawing a circle of radius 1 on the Argand diagram centred round the origin. The group then consists of all the complex numbers, $x + iy$, on the circumference of this circle. If we imagine a radius vector drawn from the origin to any point on the circumference, the length of the vector will remain at unity regardless of the angle that the vector makes with the x- or y-axis, which is equivalent to an arbitrary phase term θ between 0 and 2π. In fundamental physics, the $U(1)$ group is significant as the one connected with the inverse-square law of force, which is derived from the spherical symmetry of space round a point source, and is particularly associated with the electric interaction, which has no other component. In the quantum version of this interaction, the single generator of the group becomes the photon, the boson which mediates the interaction.

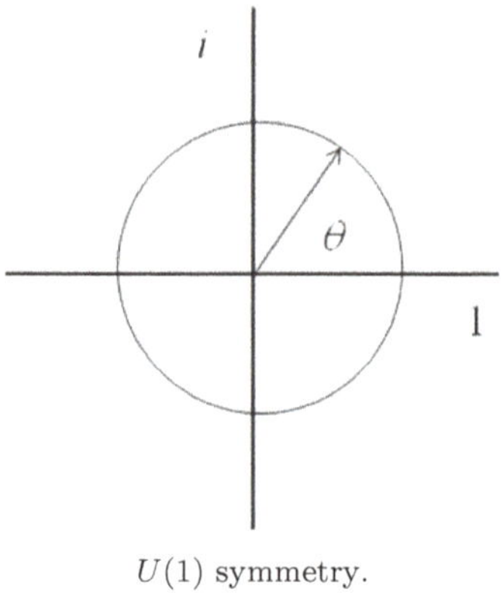

$U(1)$ symmetry.

Of particular significance in physics are the special unitary groups of degree n, or $SU(n)$, which are groups of $n \times n$ unitary matrices, like $U(n)$ groups, but this time with the extra specification that the matrices have determinant 1. $SU(n)$ groups have $n^2 - 1$ generators, the extra constraint of determinant 1 reducing the degrees of freedom, and hence the number of generators, by 1. $SU(2)$ (which is isomorphic to the group of unit quaternions and can be mapped smoothly onto the 3-sphere) is significant as the symmetry group of the weak interaction in the Standard Model, the 3 generators being equivalent to the 3 weakly interacting bosons, W^+, W^- and W^0. In the Standard Model the W^0 mixes with the $U(1)$ field to produce the Z^0 and the photon. It is also the symmetry group for fermion spin. $SU(3)$ is the symmetry group for the strong interaction, with 8 generators, which can be identified with the gluons or bosons mediating the interaction.

A more advanced discussion of Lie groups would show that those groups of most interest in physics, including $SU(2)$, $SU(3)$, and the exceptional groups F_4, E_6, E_7 and E_8, can be derived from the symmetries associated with the real numbers, complex numbers, quaternions and octonions, and are ultimately an expression of the special nature of 3-dimensional space, glimpsed in Hamilton's discovery of quaternions.

2.4 Nilpotents and idempotents

Clifford algebra allows us to create some unusual algebraic objects, ones which are square roots of zero, which we call *nilpotents*, and ones which are square roots of themselves, which we call *idempotents*. In fact, nilpotents are more strictly defined as objects for which *any* finitely repeated operation will yield zero, but we will here be concerned only with the case where the operation is squaring. Trivially, of course, 0 is both idempotent and nilpotent, while 1 is idempotent. However, we can also create nonzero nilpotents and nonunit idempotents. The main reason for this is the incorporation of terms that are anticommuting, which then removes all the cross terms in the product by mutual cancellation. Pythagoras' theorem can be structured entirely in nilpotent form. In a simple example, a Pythagorean triplet could be written in the form:

$$(i\boldsymbol{k}5 + \boldsymbol{i}4 + \boldsymbol{j}3)^2 = 5^2 - 4^2 - 3^2 = 0.$$

$(i\boldsymbol{k}5 + \boldsymbol{i}4 + \boldsymbol{j}3)$ is now a nilpotent, or square root of zero. We could even make it a little more complicated by making the 4 a multivariate vector, commutative to the quaternions in the equation:

$$(i\boldsymbol{k}5 + \boldsymbol{i}4 + \boldsymbol{j}3)^2 = 0.$$

To create an idempotent, we can premultiply $(i\boldsymbol{k}5 + \boldsymbol{i}4 + \boldsymbol{j}3)$ by $\boldsymbol{k}$ and a scaling factor a to be determined later. So we try:

$$a\boldsymbol{k}(i\boldsymbol{k}5 + \boldsymbol{i}4 + \boldsymbol{j}3)a\boldsymbol{k}(i\boldsymbol{k}5 + \boldsymbol{i}4 + \boldsymbol{j}3).$$

We can multiply this all out term by term, but there is also a shortcut, because

$$a\boldsymbol{k}(i\boldsymbol{k}5 + \boldsymbol{i}4 + \boldsymbol{j}3)\boldsymbol{k} = a(-i\boldsymbol{k}5 + \boldsymbol{i}4 + \boldsymbol{j}3)$$

and

$$(-i\boldsymbol{k}5 + \boldsymbol{i}4 + \boldsymbol{j}3) = (-2i\boldsymbol{k}5 + i\boldsymbol{k}5 + \boldsymbol{i}4 + \boldsymbol{j}3)$$

which means that

$$\begin{aligned} a\boldsymbol{k}(i\boldsymbol{k}5 + \boldsymbol{i}4 + \boldsymbol{j}3)a\boldsymbol{k}(i\boldsymbol{k}5 + \boldsymbol{i}4 + \boldsymbol{j}3) &= a(-2i\boldsymbol{k}5)a(i\boldsymbol{k}5 + \boldsymbol{i}4 + \boldsymbol{j}3) \\ &= -10ia^2\boldsymbol{k}(i\boldsymbol{k}5 + \boldsymbol{i}4 + \boldsymbol{j}3) \end{aligned}$$

So we make $a = -1/10i$ and $-\boldsymbol{k}\ (i\boldsymbol{k}5 + \boldsymbol{i}4 + \boldsymbol{j}3)/10i$ becomes an idempotent, or square root of itself.

2.5 Standard and non-standard analysis

There are one or two aspects of mathematics that will be important to us but that we don't need to cover in detail. One concerns real numbers and also calculus and the properties of space. Real numbers begin with the natural numbers, 1, 2, 3, *etc.*, which are quickly generalised to the integers, which may be negative as well as positive. The next stage is to take fractions of integers and define these as rational numbers. Then come algebraic numbers, which are solutions of algebraic equations, that is equations in which a polynomial expression in one variable is equated to 0, for example $x^2 - 2 = 0$. Finally, there are transcendental numbers, such as π and e, which are not generated by algebraic equations, but can usually be expressed only in terms of an infinite series. All of these combined are the real numbers.

Most of us are familiar with the Cantor argument by which rational and algebraic numbers can be counted by being put into a one-to-one relation with the integers, but the transcendental numbers (and hence the real numbers as a whole) cannot. In the Cantor argument, between any two numbers that can be counted in this way there are an infinite number of real numbers that cannot. The real numbers, therefore, are uncountable or non-denumerable. This argument is valid and most people think it is uniquely so, in that the opposite construction of countable real numbers is false. This, however, is not the case. Even though the Cantor argument is valid in that real numbers can be defined in this way, it is *not uniquely true*. Real numbers can be constructed in such a way that they can be put into a one-to-one correspondence with the integers and so can be made denumerable in the same way as all other numbers. This is because to construct a real number requires an algorithmic process, and algorithmic processes can be counted. So, if we think of real numbers as simply 'there' in nature, then they cannot be counted. If we think of them as always the result of a construction, then they can.

The procedure for counting real numbers in this way was first generated by Skolem in 1934 and is called 'non-standard arithmetic' (as opposed to 'standard arithmetic', based on the Cantor argument). It relates to the Löwenheim–Skolem theorem, in which any consistent finite, formal theory has a denumerable model, in which the elements of its domain are in a one-to-one correspondence with the positive integers. It was subsequently applied to the real numbers in the construction of space in non-Archimedean geometry, a fact which will be specially important to us. A parallel development also applies to calculus in which non-standard analysis aligns itself with non-standard arithmetic in the same way as standard analysis aligns itself with the arithmetic based on the Cantor continuum.

Many people will remember being taught differentiation using infinitesimals. You would draw a graph of a function and draw a line between two points on the graph with horizontal and vertical separations δx and δy, then use the function of the graph to find an expression for $\delta y/\delta x$. You would then make the two points approach each other and say that at an infinitesimal distance apart you could cancel any term with δx in it, leaving you with an expression for the differential dy/dx. This was a very effective procedure, used by Newton and others in the seventeenth century and cast into the form familiar to us by Leibniz, but it was always considered nonrigorous. To do differentiation 'properly' meant 'taking the limit', a process introduced by Newton and perfected by Cauchy, in which there were no such cancellations. However, in the twentieth century, Abraham Robinson developed a new procedure in which infinitesimals could be made as rigorous as limits (incidentally, using a nilpotent-type structure).[4] It was found that this 'non-standard' procedure and the standard analysis based on limits produced exactly the same results, and neither method was always superior to the other in proving mathematical theorems. Sometimes one method was superior, sometimes the other. Remarkably the methods of standard and non-standard analysis and arithmetic, and Archimedean and non-Archimedean geometry, are completely dual and cannot be distinguished by any known mathematical theorem or methodology.

2.6 Topology

Topology has many insights that are valuable in physics. Here, we are concerned only with one. This is the distinction between simply- and multiply-connected spaces. A simply-connected space is one without singularities. A multiply-connected space is one with a topological singularity. Now, if we imagine 'parallel-transporting' a vector round a closed circuit in the simply-connected space, that is, moving the vector round the circuit in such a way that it is always at a tangent to the path, then, on returning to the starting position it will be pointing in the original direction. If the space is multiply-connected, however, it will, on its return, be pointing in the *opposite direction*. That is, there will be a phase change of 180° or π radian. To return to the start pointing in its original direction, it will have to do a *double circuit*. This is an illustration of a general phenomenon, known as the Berry phase or geometric phase, and, although it is normally described in geometric terms, it can always be mapped onto a topological representation.

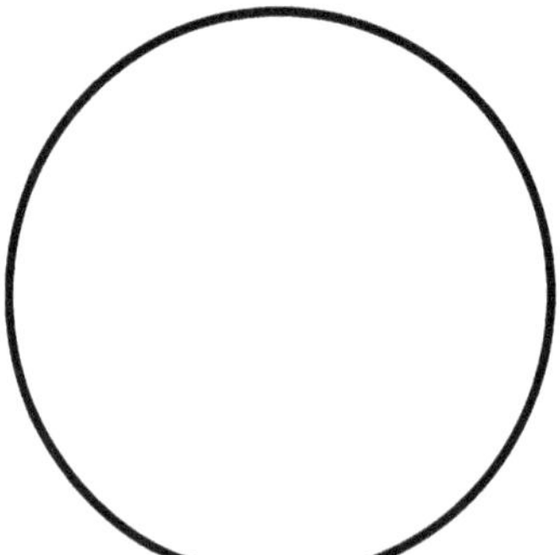

Simply-connected space.

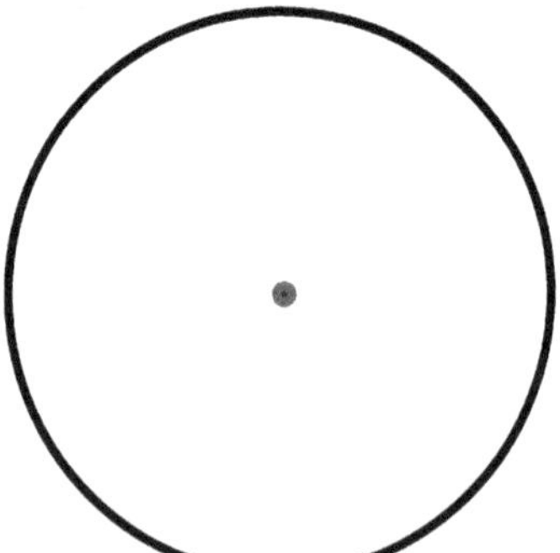

Multiply-connected space.

The topology of $SU(n)$ groups, notably, is simply-connected, while that of $U(n)$ groups is not.

2.7 Key numbers in duality, anticommutativity and symmetry-breaking

The final mathematical topic is the simplest, but it is one of the most important. At this stage it is simply introduced as a comment to be followed up at a later period. A few key integers turn up regularly in physics and sometimes in other areas of science (particularly, in biology). Some of these have a very primitive origin in the deepest levels of physics, though this has never been recognised by those who think they need 'sophisticated' explanations. The most important of these are 2, 3 and 5. Generally, where 2 occurs it can be tracked down to duality, where 3 occurs it comes from anticommutativity, and where 5 occurs it represents symmetry-breaking.[6] 5 always comes from complexity, but the other two numbers come from a genuinely primitive level. It is astonishing that nearly all the group structures that are important in physics, including such seemingly complicated ones as E_8, can be interpreted in terms of these numbers, and the individual introduction of the numbers tracked down to their roots.

2.8 Some significant group tables

Complex numbers as a group:

*	1	–1	*i*	–*i*
1	1	–1	*i*	–*i*
–1	–1	1	–*i*	*i*
i	*i*	–*i*	–1	1
–*i*	–*i*	*i*	1	–1

Quaternions as a group:

*	1	*i*	*j*	*k*	–1	–*i*	–*j*	–*k*
1	1	*i*	*j*	*k*	–1	–*i*	–*j*	–*k*
i	*i*	–1	*k*	–*j*	–*i*	1	–*k*	*j*
j	*j*	–*k*	–1	*i*	–*j*	*k*	1	–*i*
k	*k*	*j*	–*i*	–1	–*k*	–*j*	*i*	1
–1	–1	–*i*	–*j*	–*k*	1	*i*	*j*	*k*
–*i*	–*i*	1	–*k*	*j*	*i*	–1	*k*	–*j*
–*j*	–*j*	*k*	1	–*i*	*j*	–*k*	–1	*i*
–*k*	–*k*	–*j*	*i*	1	*k*	*j*	–*i*	–1

Octonion multiplication table:

*	1	i	j	k	e	f	g	h
1	1	i	j	k	e	f	g	h
i	i	-1	k	$-j$	f	$-e$	$-h$	g
j	j	$-k$	-1	i	g	h	$-e$	$-f$
k	k	j	$-i$	-1	h	$-g$	f	$-e$
e	e	$-f$	$-g$	$-h$	-1	i	j	k
f	f	e	$-h$	g	$-i$	-1	$-k$	j
g	g	h	e	$-f$	$-j$	k	-1	$-i$
h	h	$-g$	f	e	$-k$	$-j$	i	-1

Chapter 3

The Most Primitive Concepts

3.1 What are the most primitive concepts in physics?

This chapter and the next will be the most important in the whole book. Here, we will begin to justify the claim that the foundations of physics is a subject in its own right and with its own methodology. What we are going to do will be almost entirely inductive, but also deeply mathematical. It won't yet look like conventional mathematical physics because that is a product of complexity and emergence, not of fundamental simplicities, but it will certainly lead to it, and we will show how in the chapters that follow. We have established our methodology and laid out our mathematical toolkit. Now we will apply these to physics at its most fundamental, 'embryonic' level. If our application is correct, then physics can never be looked at in the same way again. Although we will be using the methodology generated by our philosophical reasoning, this work will not itself be an exercise in philosophy, but a completely physical discussion pitched at the most primitive level at which physical concepts can be identified as such. Questions that cannot be answered in physics at a more developed, emergent level, will, we expect, find answers when pitched at this foundational level.

When we are working at this level, the inductive mode has to go into overdrive. We need entirely new techniques of organizing the information that is confronting us in every direction. We have to use our methodological principles and the pre-prepared mathematics to see past all the conflicting claims and the layers of complication that nature, as well as ourselves, has put in the way of getting directly to the central core of information. We need to develop an instinctive feel for the 'big picture' and be able to sort out the ideas that are truly basic from the accumulation of complexity that is beginning to interfere with them. The big simple ideas, like why time flows only one way, need big simple answers — not ones based on the emergent consequences, as we may be tempted to think — and we have to develop an aptitude for recognising the recurring patterns. But this inductive thinking is very different from pure speculation. There should be no constraints on the ideas we generate, but there are, and should be, very definite and very strong constraints on the ones we *accept*, because the methodology allows us immediately to work out the consequences that a *foundational* idea will generate. Because they have such a general application, foundational ideas that are wrong will immediately generate a mass of unacceptable consequences. Ones that are broadly correct will tend to produce smaller issues, which successive iterations are likely to resolve. The special advantage of working in a purely abstract mode and avoiding model-dependent ideas is that faults in the structure can't be hidden in the details, as they can with a model. To a large extent, it either works or it doesn't, and the main problems arise with deciding which ideas are more basic than others.

Near the end of the first chapter we outlined those concepts that we thought the most 'primitive', based on those that survived as we changed our scale of operation. These included space, time, something representing matter in its point-like state and something representing energy or the connections between the points of matter. We can now approach the last two concepts more directly for it is clear that they, in some sense, represent the sources of the four known physical interactions, which, apart from space and time, seem to be all that could be truly fundamental in physics. The concept related to the point-like nature of matter can be identified as charge. However, this is not as simple as we might think. Apart from electric charge, the weak and strong interactions require a similar concept, and the parallel is certainly assumed in the concept of 'charge conjugation'. In fact, in view of the already successful partial unification involved in the electroweak theory, and the potential for further unification at higher energy scales, it seems more meaningful to define a single parameter with three components than to imagine that there are three totally separate concepts.

However, something seems to break the symmetry between the three charge components that we might otherwise expect in such a picture, and that has been assumed to exist at some particular energy regime in Grand Unification, and is further suggested by the common $U(1)$ component that each of the forces has. A situation like this needs to be attacked using a bold conjecture (in fact, at this level we have no alternative), based perhaps on the idea that we may have seen the pattern before. In fact, we have seen in our mathematical discussion where we have two 3-dimensional structures how 'packaging' can lead to a broken symmetry. Since an 'embryo' theory must be expected to show differences from a fully-fledged emergent one, we may reasonably take it as a provisional assumption that this is what is happening here. If this is a correct assumption, and, as far as it goes, it is perfectly compatible with Standard Model physics as we now know it, then the exact mathematical structure for symmetry-breaking which we have established in the previous chapter should be reflected in the physics that emerges, and we will investigate this in the next chapter. The point is that, if the conjecture is correct, the structures that separate the interactions should begin to emerge from the mathematical pattern we have imagined might create them. If they don't, it will soon become obvious.

The other concept that might be primitive has an established and single structure. It is the source of gravitational interaction, and has to be differentiated from the other three. It can be called energy, mass-energy or mass. Though some people like to reserve the last term for the invariant or rest mass of material particles, I prefer to use it for the last primitive quantity, rather than energy. One reason is to emphasize its role as a source for gravity; and I want to show later how the concept of 'energy' in the quantum sense and its relation to mass seen as the gravitational source are emergent properties from the packaging of the more primitive structure. In any case, there is no physical system in which the rest mass can actually be separated from some dynamic component. All mass is, in fact, in some sense dynamic, and, in the next chapter, we will see that even this so-called rest mass can be seen as a product of the subtle quantum dynamical process known as *zitterbewegung*, in the sense that the frequency of the latter, $\approx 2mc^2/\hbar$, is totally determined by the mass and so can be seen as a measure of it.

The concepts identified here could almost have been derived by the old method of dimensional analysis and they feature at a fundamental level in such a deeply significant result as the CPT theorem. They should be seen as purely abstract — none, as we will show, is any more 'real' than any other. They can all be seen almost as manifestations of pure algebra. Also, we are not assuming that these concepts are the most fundamental imaginable, but the most fundamental imaginable that can be described as purely *physical*. The first chapter in my book, *Zero to Infinity*, in fact, provides a kind of information process by which they and their mathematical structures can be generated from zero, and which plays a more general role in all self-organizing processes in nature. However, this is physics at a level where it can no longer be distinguished from mathematics.

3.2 Measurement

The most primitive abstract concepts must be more fundamental than either the laws of physics or the structure of matter, for neither of these can be imagined without them. It must be significant that of all the four concepts we have suggested may be primitive, only one, space, is actually susceptible to direct measurement, even though measurement is seemingly the only means we have of investigating nature. We have created many ingenious devices for measuring and recording data, but however sophisticated the system, they can always be reduced to the equivalent of moving a pointer across a scale.

For some reason, we have found it necessary to channel our measurements of space through three other conceptual structures as well, as though they had an independent existence. The entertainment industry is totally dependent on the fact that attributes of space, such as shape, colour and sound vibration, can be used to simulate things that are not meant to be spatial. The make-believe of films, sound recordings and holograms, for example, depends entirely on the idea that variations in space can induce in the viewer or listener a sense of the passage of time or the presence of matter. In fact, space is not only unique in making measurement possible, it is also universal in that any object or collection of objects will automatically create a measurement standard for space, and this will happen at all times and in all places throughout the universe.

Many people will think that we can also measure time, but, in fact, though we can sense time passing through the laws of thermodynamics, and can detect whether the simulation of a sequence of events is running in the wrong direction, we can never actually measure it. When we claim to be 'measuring' time, we are really only measuring space. Of course, it's a bit different from measuring space in the ordinary way, because it has to be repetitive. We have to find some kind of device in which something repeatedly traverses the same section of space, so that we can construct a time interval from the frequency of repetition. We rely here on the supposition or realisation that time only flows in one direction while the space direction is repeatedly reversed, so time is inferred by measuring the total distance travelled as a scalar, rather than vector, quantity. Traditional devices, such as pendulum clocks and watches governed by a balance spring, worked on the principle that simple harmonic motion was *isochronous*, or regular in its repeating cycle. The alternative was to use an astronomical measure, like the rotation of the Earth or the orbit of the Earth round the Sun.

Relatively modern devices like atomic and digital clocks work on the same principle, with internal oscillations which are counted automatically. But in all the devices the space is observed not the time.

Clearly, we need special conditions to 'measure' time, and these would be impossible without acceleration and force. Even sending a light signal over a known path requires a reflection, and so still uses the same principles of force and repetition. Acceleration and force are second order in time and so are yet further removed from anything like a direct time measurement. The same is true of 'measurements' of mass and charge, which again are only possible in the presence of force, and again require the equivalent of a pointer moving over a scale. The mass of solid objects is determined by observing the force of gravity on them at the Earth's surface, and this involves both a spatial measurement and the use of a clock and its spatial repetitions. The same is true for astronomical objects, where we observe the dynamics due to gravity on a large scale. There are various ways of measuring the mass-energy of a particle, but they all involve using a force or a heating effect, with consequent reliance on spatial observation. Charge, of course, is not even detectable without force.

Now, the special nature of space has long been of interest to philosophers and scientists, and several have tried to reduce the whole of nature to this one quantity. Descartes believed that there was nothing other than extension in nature and that matter had no separate existence. Einstein, of course, set out to build a physics in which there was nothing but space. Time became the fourth dimension of the new combined concept of 'space-time' in the special theory of relativity, while mass-energy was expressed as space-time curvature in the general theory. He never succeeded in incorporating electromagnetism, which Kaluza and Klein decided needed the addition of a fifth dimension. No one has come close to incorporating the rest of the Standard Model into a space-like structure even by incorporating yet more dimensions, though this appears to be the ultimate aim of string theory.

But even with such 'unifications' as we have already achieved, there appear to be significant problems. A unification which created a single multidimensional space-like structure would not fulfil our criterion of leaving nothing arbitrary in a fundamental theory. Space itself, and dimensionality, would remain unexplained, and there would also be no explanation of why the dimensions had such distinct manifestations. If the multidimensional 'space-time' really was the foundation for physics, there would be nothing more foundational to explain its components. In any case, quantum mechanics, which is by far the most successful physical theory ever devised, seems to be telling us that space and time are fundamentally different in ways that suggest that a theory relying on a *complete* union, while leading to some interesting consequences, will fail at some significant point. One of the most significant differences in quantum mechanics is that space is an observable, while time is not, exactly in line with our more general analysis. The indications are that the union between space and time in relativity, is an emergent one, at the first stage of complexity, not a foundational one that can't be broken down further.

It is surely significant that, although space is the only observable and measurable quantity, nature seems to be telling us that we also need time, mass and charge, concepts whose relationship with space is anything but direct. There have been numerous attempts, especially in the twentieth century, to reconstruct physics as an observer-centred subject, in which the only parameters appearing in physical theories should be those that are directly observable, and quantum mechanics originated from one such attempt. However, it quickly became clear that quantum mechanics also required quantities that could not be observed, or could not be successfully observed at the same time as others. The main effect was to change *which* quantities were to be classified as observables; it was unable to specify that observables must be exclusive. Ultimately, this is what we must expect if we are true to the principles outlined in the first chapter. Measurement and observability, however desirable, are no more likely to be universal aspects of a nature that cannot be absolutely characterised than anything else.

Nevertheless, there is clearly something special in the fact that we can get so near in *almost* reducing the properties of the other parameters to those of space, even though the attempt gets increasingly problematic as we work through the connections with time, then mass, and finally charge. There has to be a connection which shows why space becomes a 'privileged' concept to which these others so nearly relate, and we will return to this later in the book with a quite unexpected solution.

Our main task now is to apply the principles we have established to examining the structures of the parameters in relation to each other. At first, this may appear low key, but the comparisons on a fundamental basis will show a progressive tightening of the options available and a progressive tightening of the descriptions, which will eventually lead to an understanding from which there will be no going back. Once we have reached this stage, it will become apparent that this process, which you can imagine having gone through many previous iterations, has begun to explain even the most seemingly inexplicable fundamental questions within a framework that is mathematically rigorous, though we haven't yet needed to introduce a single mathematical equation. The following three sections will be the most significant in the entire book in showing that the foundations of physics methodology has a power to construct a basis from which more sophisticated aspects of physics can emerge at the first level of complexity.

3.3 Conservation and nonconservation

Though we have been unable to reduce all physical concepts to aspects of space, we remain convinced that there must be *relationships* between space and the other fundamental concepts. If we can't establish *identity* between concepts, there is an alternative, which physics and our methodology allow us, to establish *symmetries* between them. At this level, we could guess that the most likely symmetry to be found is the most basic one, duality, represented by the C_2 group. If we had perfect duality between two things, we would expect to find some characteristics in which they were identical, and some in which they were absolutely opposite. If we had exactly dual concepts, then they might well look alike in many, even most, respects, and we might even believe for a long time, that they were identical. It might then take a lot more searching before we found the areas where they were different.

Perhaps this is the position we have now reached with space and time, or even space and all the other quantities. Possibly, we should consider them as dual, rather than identical. In fact, for a fundamental theory, duality would offer a better route to explanation than identity, for, if all other concepts were versions of space, then we would have no route to finding an explanation of space itself, whereas dualities might only be explicable if we could explain the concepts themselves. So, do we have any dualities at the foundations of physics? This and the next two sections will explore three possible cases, the first being that between conservation and nonconservation.

If we were asked to guess which laws of physics might be absolutely true, a good bet would be the conservation laws. Most descriptions of physical systems seem to involve a statement, direct or indirect, that some quantity is conserved — for example, mass-energy, momentum, angular momentum, charge — while others — for example, space and time — are not. Noticeably, these are largely concerned with our fundamental parameters. Now the conservation laws of mass (or mass-energy) and charge are among the most fundamental in the whole of physics. They are also very specific. Mass and charge are not just globally, but *locally* conserved, that is a point-charge or an element of mass is conserved at a point in space and time and cannot be destroyed at one point in space and time to be recreated in another. It is as though each elementary charge and each element of mass had an *identity*, or unique label, which it carried with it throughout any changes brought about by its interactions, except in so far as a charge of one sign can be destroyed by one of the opposite sign.

What about quantities that are not conserved? Is there a property that can be called 'nonconservation'? Surprisingly, there is and it is just as definite a property as conservation, and an exact dual to it. This is the property of the quantities we call variable, in particular, space and time, quantities which have *no identity*. We can't single out a unit of space and time, like we can those of mass and charge, and there are three major symmetries which say exactly that. According to the translational symmetry of time, one moment in time is the same as any other. There is no way of pinning down a moment. We can 'translate' or move linearly along the time direction without any noticeable effect. There is, similarly, a translational symmetry of space which says that one element of space is the same as any other, and that absolute position in space is arbitrary. Space, however, is also a 3-dimensional quantity, and this leads to a third, rotational, symmetry, which says that one direction in space is the same as any other. The translational and rotational symmetries of space combine to produce its affine structure, which allows us to reconstruct a vector in space along axes in any direction.

If translational and rotational symmetry are properties of nonconserved quantities, the conserved quantities should have exactly opposite properties, and they do. Mass and charge are both translationally *a*symmetric and we can guess that charge, if it is truly a 3-dimensional quantity, is additionally rotationally *a*symmetric. The translational asymmetry is obvious: a unit that is unique cannot be replaced by another. This is clearly true of charges. Here, we note that even though quantum mechanics says that the *wavefunctions* of identical fundamental particles are 'indistinguishable', this is a property of the *observed* space and time aspects, not of the charges. The same is true of the positioning: wavefunctions are extended, charges are not. For mass-energy, though the element may undergo a continual transformation of form, it nevertheless retains its identity. (The 'identity of energy' was a remarkable contribution made by Oliver Lodge to the explanation of the Poynting theorem on the flow of electromagnetic energy.[5])

The power of the foundational method now becomes apparent when we ask if charge, as a 3-component or 3-'dimensional' quantity, shows any property which we might reasonably interpret as rotational asymmetry. Experimental evidence so far suggests very strongly that it does. The three types of 'charge' (electric, strong and weak) do not 'rotate' into each other, despite the partial unification of the electric and weak forces in the $SU(2) \times U(1)$ electroweak theory. They must be separately conserved. Now, in particle physics, the composite baryons, such as the proton and neutron, are the only particles known with net strong 'charge', which manifests itself as 'baryon number'. If strong charges are conserved separately from electric and weak charges, then there is no end product for a baryonic decay except another baryon.

Baryons, also, along with leptons, are classified as fermions. Essentially, fermions, as distinct from bosons, are particles which are sources of the weak interaction and have net weak charges. (The W and Z bosons are carriers of the weak interaction, but not sources of it — the same applies to photons with respect to the electric interaction.) Leptons cannot decay into particles which have no net weak charge, and cannot decay into baryons, which have net strong charges, so leptons can only decay into other leptons. Experimental results have repeatedly shown that both baryon and lepton numbers are conserved in all particle interactions. Though there have been repeated speculations that protons could decay in such a way as to violate these laws, for example into a neutral pion (with no charges) and an antielectron (with just electric and weak charges), experimental results have always contradicted them, and the limits on the proton lifetime have been extended way beyond the predictions originally made for decay on the basis of grand unification.

Another important property of nonconserved quantities, which comes very close to translational/rotational symmetry, is *gauge invariance*, which we see in both classical and quantum contexts. A quantity that is invariant remains constant when everything that can change does so. To define a conservative system, we need to define it as one in which the conservation laws apply. The universe is such a system and so are individual particles, but there are also many systems in between where they apply to a very good approximation, and physics generally operates by constructing such systems. Under the principle of gauge invariance, the field terms which determine the strength of interactions are unchanged even when there are arbitrary changes in quantities that are subject to translations (or rotations) in the space and time coordinates, such as the vector and scalar potentials, or phase changes in the quantum mechanical wavefunction. In effect, a conservative system can absorb arbitrary changes in the space and time coordinates as long as there are no changes in the values of conserved quantities, such as charge, energy, momentum and angular momentum.

From classical physics, we have the examples of the scalar electric and gravitational potentials, which are each defined, in principle, as the ratio of the source (charge or mass) to the distance from the source. The quantity will vary with the zero point from which we measure the distance. However, it is actually the potential *difference* between two points which determines the energy transfer between them, and, as long as the charge or mass value is conserved, the zero position is irrelevant. In quantum terms, changing the absolute values of the nonconserved quantities, space and time, is equivalent to changing the arbitrary phase involved in the interaction, an expression of the $U(1)$ symmetry involved in defining the interaction sources in terms of scalar potentials. The absolute value of the phase term in an interaction remains unknown because it has no effect on any physical outcome. We can thus divide fundamental physical quantities into those whose absolute values are significant (the conserved ones) and those where they are irrelevant (the nonconserved ones). Significantly, in the Yang–Mills theories, which govern particle interactions in the Standard Model, gauge invariance is *local*, exactly like the conservation laws. The property of local conservation of charge and mass leads to an exactly opposite dual property of local nonconservation of space and time.

Now, gauge invariance and translational and rotational symmetry are not merely passive constraints. They force us to construct physical equations in such a way that nonconserved quantities have properties exactly opposite to those of the conserved ones, and that it is *explicitly shown* that this is the case. So, we write the laws of physics in terms of *differential equations*, with the nonconserved or variable quantities expressed only in terms of the rates of change, which specify that they are not fixed. The differential equations ensure that the conserved quantities — mass and charge (and also others derived from them, such as energy, momentum and angular momentum) — remain unchanged while space and time, expressed as the differentials, dx, dt, vary absolutely. It is not enough to say that space and time have no fixed values. They must be *seen* to have none.

The intrinsic variability or nonconservation of space and time can be seen as the ultimate origin of the path-integral approach to quantum mechanics, where we must sum over all possible paths. None can be privileged. It is also responsible for many of the aspects of quantum mechanics that seem to concern the naïve realists. The fundamental meaning of nonconservation seems to require that God does 'play dice'. So we have to accept that space and time, as nonconserved quantities, are in principle subject to absolute variation, as long as they do not violate conservation principles. It is the conservation principles alone that restrict the range of variation of space and time when particles and systems interact. On a large scale, with many such principles acting simultaneously, the variation can be reduced to the point where we can make a classical 'measurement'. But it is not the *measurement* that makes the situation become classical. The degree of variability, in fact, becomes restricted by the application of external potentials, requiring new conservation conditions, as the isolated system interacts with its external environment (the 'rest of the universe'). The so-called 'collapse of the wavefunction' is nothing more significant than the extension of an isolated quantum system to incorporate some part of its environment, so introducing a degree of decoherence.

A free electron has complete variability and can be anywhere at any time. It remains free of any conservation principle that would restrict it (except the conservation of charge). If we now bring it near a proton, constructing a hydrogen atom, we find that it is now subject to new principles of conservation of energy and angular momentum which apply to the system. Nevertheless, the electron can still be anywhere at any time, as long as those principles are obeyed. The electron's position cannot be fixed, but its range of variability is no longer the whole of space, but rather that determined by the conservation principles. If we now bring our hydrogen near to another one to form a hydrogen molecule, the electron's position is still not fixed, but its range of variability has been changed according to the new conservation of energy and angular momentum principles which apply to the hydrogen molecule. Eventually, we can extend the system to the point where the variability is below anything we can observe.

It is clear that a great deal can be learned about both the conserved and the nonconserved parameters by realising that the distinction comes from a dual pairing. There is even a well-known mathematical result which is an expression of the duality. According to Noether's theorem, every variational (i.e. variable) property in physics leads to a conserved quantity.

Three classic examples of this relate to the translation-rotation properties we have already discussed.

translational symmetry of time	≡	conservational of energy
translational symmetry of space	≡	conservation of momentum
rotational symmetry of space	≡	conservation of angular momentum

According to our terminology, and the mass–energy relation $E = mc^2$, the first of these suggests that translational symmetry of time also demands the conservation of mass. This is exactly what we would expect if non-conservation and conservation were exactly dual, as applied to time and mass. If our general methodological principles are true, this would lead us to expect a corresponding link between nonconserved space and conserved charge. In fact, as these are assumed to be 3-dimensional quantities, the link would manifest itself in two different ways, referring to translation and rotation. So we should expect:

translational symmetry of time	≡	conservation of mass
translational symmetry of space	≡	conservation of value of charge
rotational symmetry of space	≡	conservation of type of charge

Is the link true? Work done as long ago as 1927 by Fritz London suggests that the translation part, at least, might be valid. In fact, it is almost obvious from the principle of gauge invariance. London showed that the conservation of electric charge was identical to 'invariance under transformations of electrostatic potential by a constant representing changes of phase', the phase changes being of the same kind as those involved in the conservation of momentum. In principle, as in *gauge invariance*, the electric charge is conserved while the scalar electric potential (or ratio of charge to distance from the source) is not. In fact, we can extend the result beyond the electric charge, for the strong and weak charges each have an associated potential of the same kind as the electric one (a Coulomb potential). This potential essentially determines the value of the coupling constant for the interaction and so can be said to determine the 'value of charge'. So, it looks like this result will fulfil the test.

What about the other? This is a major development. The foundational method proposes a result which is completely unprecedented, and, frankly, looks bizarre. How can we make the conservation of angular momentum relate to the conservation of *type* of charge? How can it show that there is no mutual transformation between weak, strong and electric charges, and that the laws of baryon and lepton conservation will therefore hold? It looks impossible, but there is, in fact, an extraordinarily simple explanation which relates to the broken symmetry between these forces in the Standard Model. It provides the first significant test of our whole approach. However, we have to establish a few other things before we can give the complete explanation.

3.4 Real and imaginary

The next area of investigations brings in the algebras we have considered in the previous chapter, and especially the norm 1 and norm −1 units that make up Clifford algebra. We will use 'real' to indicate norm 1 quantities, those whose units square to 1, and 'imaginary' to represent the norm −1 quantities, whose units square to −1, whether or not they are vector or scalar, commutative or anticommutative. The question of whether quantities are real or imaginary is a very significant one in physics because squaring occurs in nearly all aspects of the subject, for example, in Pythagorean addition for space and time, the amplitudes in quantum mechanics, and even in the interactions of masses and charges.

Imaginary numbers have been important to quantum mechanics from the beginning, in addition to noncommutative algebras, but, even in classical physics, it was obvious from Euler's theorem relating sines and cosines to exponentials that the mathematics of waveforms was greatly simplified by the use of complex numbers. The introduction of Minkowski space-time for relativity led to the development of 4-vectors, or quantities with 3 real parts and one imaginary, almost as Hamilton had imagined in the early days of quaternions. So, from the representation of space and time as a version of Pythagoras' theorem in 4D,

$$r^2 = x^2 + y^2 + z^2 - c^2t^2 = x^2 + y^2 + z^2 + i^2c^2t^2$$

where r^2 is a scalar product, we extract the 4-vector

$$r = \mathbf{i}x + \mathbf{j}y + \mathbf{k}z + ict.$$

And from the Einstein energy-momentum relation (with $c = 1$)

$$m^2 = E^2 - p_x^2 - p_y^2 - p_z^2 = -i^2E^2 - p_x^2 - p_y^2 - p_z^2$$
$$p_x^2 + p_y^2 + p_z^2 - E^2 = p_x^2 + p_y^2 + p_z^2 + i^2E^2$$

we can extract the 4-vector

$$\mathbf{i}p_x + \mathbf{j}p_y + \mathbf{k}p_z + iE,$$

which, with the c terms included, becomes

$$\mathbf{i}p_x c + \mathbf{j}p_y c + \mathbf{k}p_z c + iE.$$

Why is the time or energy component imaginary compared to the space or momentum? Why do they have different norms? If we want a *physical* argument based on relativity, it is because the light signal is retarded. But we have to remember that, at the foundational level, there is no light and there is no relativity. Sometimes, people describe the 3 + 1 real-imaginary representation of space and time as a mathematical 'trick', but mathematical tricks only work if they are needed. We have seen that vectors in our representation require an imaginary fourth component (a pseudoscalar) because they derive from complexified quaternions. There seems, then, to be a possible mathematical explanation for the representation, but there is also a *physical* one.

Physics consistently tells us that quantities containing time to the first power, such as uniform velocity, have no real significance or physical meaning. This only comes when they incorporate time squared, as we find with acceleration and force. We have already seen that time 'measurement' requires these quantities, even for time-measuring devices that use light itself. This is totally consistent with what we might expect for an imaginary quantity, but there is yet another physical reason. One thing that we haven't yet discussed about imaginary quantities is that they are intrinsically dual. They have + and – solutions which can't be distinguished. It's not that we can accept one value and discard the other — we have to always include both. Any equation with a 'positive' imaginary term in its solution has to have a dual solution with 'negative' imaginary terms. If we are using imaginary numbers, then we are automatically accepting the duality that is built into them. So, we have no option if we use it or ict but to regard it as implying a duality in its sign. Now, one of the best known aspects of time is that it flows only one way, a fact that we can detect from the increased entropy or disorder that follows any physical event. Physical equations, however, seem to ignore this completely and are constructed to have two directions of *time symmetry*. Even if we can't reverse time, we can extract physical meaning from reversing the sign of the time parameter (as with *CP* violation in particle physics). This constitutes the famous reversibility paradox. However, there is no paradox at all if time really is imaginary, and the one-way flow of time comes from an entirely different aspect of the parameter (as we will see in the next section). A parallel case can be seen in relativistic quantum mechanics where two signs of the energy parameter derive from time via its representation as $\partial/\partial t$, though there is only one sign of physical energy.

If our methodology is correct, then we might expect to find a real-imaginary distinction occurring also with mass and charge. Of course, we have the problem that our picture of 'charge' is complicated by the fact that there is a broken symmetry involved, which ensures significant distinctions between all three interactions for which we suppose it is the source. On such occasions, I tend to assume that a real unbroken symmetry is there which is exact in principle, and that the breaking of the symmetry is an effect of emergence or complexity, as our analysis of the Clifford algebra would suggest. Subsequent chapters will show how this assumption can be completely justified.

It is widely believed that there is some energy regime at which the weak, strong and electric interactions would lose their distinguishing features and become alike, but they are already alike in at least one aspect. That is, that they have a 'Coulomb' or inverse square force term, representing the $U(1)$ symmetry of a scalar phase. In this, they are also like Newtonian gravity (an aspect which is imported even into general relativity as the Newtonian potential). The weak and strong interactions differ from the electric interaction in having additional terms in their force laws which give them additional properties. It is possible then that, at Grand Unification, these extra components could be seen to shrink, leaving all three interactions as purely Coulomb in form.

Now, the inverse square force or Coulomb force has a relatively simple explanation. It is the exact result we would expect for a charged point source in a 3-dimensional space with spherical symmetry. In all known interactions, it relates to the coupling constant. This is not strictly the charge, which is really just a pure number which indicates whether or not a particular source of one of the interactions is present or not, but one can define the *magnitude* of the charge (electric, weak or strong) in terms of the electromagnetic, weak or strong coupling constants. In quantum terms, the coupling constant squared becomes the probability of absorbing or emitting the boson that carries the interaction, and this will be zero if the particular charge is not present.

One of the big unanswered questions of physics has always been why particles with identical masses attract, whereas particles with identical charges of any kind repel. This is seen clearly if we write down the force laws for the gravitational force between masses m_1 and m_2 and the electric force between charges e_1 and e_2 over the same distance r. The force for gravity is negative, which signifies attraction, meaning that the force has the opposite direction to the space vector $\mathbf{r}$. However, the electric force is positive, signifying repulsion, or a force in the same direction as $\mathbf{r}$.

$$F = -\text{constant} \times \frac{m_1 m_2}{r^2}$$

$$F = \text{constant} \times \frac{e_1 e_2}{r^2}$$

The force between e_1 and e_2 will only be attractive if they have opposite signs. We can, however, find an immediate solution if we suppose that the charges are imaginary, say ie_1 and ie_2. The force laws then assume an identical form.

$$F = -\text{constant} \times \frac{m_1 m_2}{r^2}$$

$$F = -\text{constant} \times \frac{ie_1 ie_2}{r^2}$$

We are almost drawn to this solution from the knowledge that we need to accommodate three 'charges' with the same property, but quite distinct from each other, and that the mathematics for this is readily available in the form of a quaternion, with components such as $\boldsymbol{i}s$, $\boldsymbol{j}e$, $\boldsymbol{k}w$, where s, e and w are the strong, electric and weak charges. In this context, we remember that a quaternion needs a real fourth term, or scalar, just as our multivariate 4-vectors needed a pseudoscalar, and that mass is available to play this part. We can then propose that the units of charge and mass could act as the three imaginary plus one real parts of a quaternion, just as the units of space and time act as the three real and one imaginary parts of a multivariate 4-vector.

space	time	charge	mass
$\mathbf{i}x\ \mathbf{j}y\ \mathbf{k}z$	it	$\boldsymbol{i}s\ \boldsymbol{j}e\ \boldsymbol{k}w$	$1m$

Perhaps this could be the final vindication of Hamilton, giving quaternions a direct role in nature, as well as the indirect one of being the progenitor of the 4-vectors linking space and time. In fact the quaternion representation would be logically 'prior', not only in the mathematical sense, but also in the physical sense, the character of space-time being predetermined by the necessity of symmetry with charge-mass, whose structure is completely determined by the quaternionic form. In addition, such a symmetry would constrain the vector character of space to the extended form required from a complexified quaternion or Clifford algebra, and not the restricted form of the Gibbs–Heaviside algebra, meaning that spin would be automatically factored into the structure of space, and not be an unexplained additional extra brought in with quantum mechanics.

This all depends on choosing charge to be the imaginary quantity, rather than mass. So, could we have chosen mass to be imaginary instead of charge? To show that we could not, we return to the fundamental property of imaginary numbers that we discussed in relation to time: they can only exist as a dual pair, with both + and – signs. This is true of extended imaginary numbers, such as quaternions, as much of ordinary complex numbers. All the indications are that mass, as we know it, has only one sign. Whether we call it positive or negative, there is only one version. Mass is 'unipolar'. The case is quite different with charge. There, we always have both + and – versions, whether the charges are electric, strong or weak.

This is the explanation of 'antimatter'. For every particle with a charge structure of any kind, there has to be a particle with charges of the opposite sign. We even call the switching of particle and antiparticle by the name of *charge conjugation*, and we are fully aware that it isn't just about particles with *electric* charge. Neutrons, which have no electric charge, have antiparticles, because the neutron still has strong and weak charges, and the antiparticle requires these to take the opposite signs. The only exception comes with particles like the photon, which have a totally zero charge structure, and are *their own* antiparticles, with only the spins reversed. Significantly, for all antiparticles, the masses are exactly the same as those of the respective particles, emphasizing once again the intrinsic unipolarity of mass by comparison with that of charge.

As with time, there is also another reason why charge must be the imaginary quantity while mass remains real. In effect, we can observe space directly by observation or measurement, or through its squared value, in Pythagoras' theorem or vector addition, whereas time can only be apprehended through its squared value in force or acceleration. Similarly, we can apprehend mass physically in two different ways, either directly or through its squared value. The direct method is through inertia or force = mass × acceleration. The apprehension via the squared quantity is through gravitation. Through inertia, we can apprehend a mass even if no other mass is present, though we need at least two masses for gravitation. In the case of charge, only apprehension through the squared quantity is available to us, via Coulomb's law; we can never apprehend an imaginary quantity like charge unless another one is present to create a real effect. A photon, for example, can only interact with a charge if it has already been radiated by another. Ultimately, in the case of both time and charge, the imaginary status is not a mathematical convention; it represents a real physical property.

3.5 Commutative and anticommutative

Already our methodology is forcing certain constraints on the way we view the fundamental parameters. We have established that there are at least two dualities which connect them, and there is, in fact, a third. We have already specified that mass and time are, respectively, scalar and pseudoscalar, which makes them commutative, while space, as a vector, must be anticommutative. If charge is correctly described by a quaternionic structure, then this must be anticommutative as well. Anticommutativity requires a quantity to be both dimensional and specifically 3-dimensional, and the reverse argument is also true. Commutative quantities, like time and mass, must be nondimensional, or, as it is sometimes termed, one-dimensional.

In addition to dimensionality, there is another very significant consequence of anticommutativity. This is the idea of discreteness or discontinuity introduced with the fact that the three components of an anticommutative system are very much like the components of a closed discrete set. Can we now further postulate that anticommutative or dimensional quantities are necessarily also discrete or divisible, while commutative or nondimensional quantities are correspondingly always continuous or indivisible? It seems odd to imagine that the only discrete quantities in physics are 3-dimensional, but that seems to be the logic of what is happening, and it would be yet another remarkable consequence of Hamilton's original discovery.

Continuous quantities clearly cannot be dimensional, because a dimensional system cannot be conceived without an origin, a zero or crossover point, and this is incompatible with continuity. Also, a one-dimensional quantity cannot be measured, because scaling requires crossover points into another dimension. Though it is often claimed that a point in space has zero dimensions, a line one dimension and an area two dimensions, this is actually impossible because a single dimension cannot generate structure. A line can only be seen as a one-dimensional structure, within a two-dimensional world, which itself can only exist in a three-dimensional one, because a two-dimensional mathematical structure always necessarily generates a third.

But, does this mean, by the counter-argument, that space, as a 3-dimensional quantity, is necessarily also discrete? The answer has to be yes. If space weren't discrete we couldn't observe it. The whole of our measuring process is based on the fact that space is fundamentally discrete. In the past, this has caused confusion, due to a fundamental misunderstanding about the nature of real numbers, and the lack of a methodology which could separate the different aspects of the fundamental parameters into primitive properties. We know, for certain, that *charge*, which we think may also be 3-dimensional, is certainly discrete, because it comes in fixed point-like units or 'singularities', which are easily countable. But space is nothing like this. Mathematicians frequently represent it, or one of its dimensions, by a real number line, which gives the appearance of being continuous. However, space, unlike charge, is a nonconserved quantity, and its units cannot be fixed. Its discreteness is one that is endlessly reconstructed. The real number line is not absolutely continuous, it is *infinitely divisible.* It presents exactly the characteristics required by non-standard arithmetic and non-Archimedean geometry. It is made up of real numbers constructed by an algorithmic process, and so is necessarily countable. If it were not, then measurement, and dimensionality, would be impossible.

There is, as far as we know, a very deep distinction between space and time, beyond any consideration of their mathematical nature as real and imaginary quantities. Time cannot be split into dimensions in the same way as space, which means that it cannot be discrete and must be continuous. The physical consequences are very significant, and allow us to complete the solution of the reversibility paradox. As a 'nondimensional' and continuous quantity, time is necessarily irreversible. To reverse time, we would have to create a discontinuity or zero-point. In addition, as a nondiscrete quantity, time can never be observed, which is exactly what quantum mechanics tells us. Observability always requires discreteness. The lack of observability is the exact reason why we treat time as the *independent variable*, by comparison with space. We write dx/dt, in fundamental equations, not dt/dx, because time varies independently of our measurements, represented by dx, which respond in turn to the unmeasurable variation in time.

Unlike space, which is infinitely divisible, time is *absolutely continuous*, that is not divisible at all. The two conditions are mathematically opposite, about as different as any physical or mathematical properties could conceivably be, though a fundamental duality in nature allows one to be substituted for the other. Clocks, as we have seen, do not measure time, but a space with which it has an indirect relation; the divisions that we measure are those of the space which is repeatedly traversed, and they require the fact that space, as a dimensional quantity can be reversed because we can define it to have an origin. They also, very often, use the fact that space has more than one dimension.

The absolute continuity of time gives us a complete explanation of a very old paradox, one of the oldest in physics. This is the famous paradox of Achilles and the tortoise, due to Zeno of Elea. In this paradox, Achilles can run ten times faster than the tortoise, so, over a hundred metre race, the tortoise gets a start of ten metres. While Achilles runs ten metres to catch up, taking one second, the tortoise runs one metre. Achilles then runs another metre while the tortoise runs a tenth of a metre. If Achilles runs this tenth of a metre, the tortoise will run a hundredth of a metre and so still be ahead. Achilles may be ten times faster, but he will never actually catch up. Zeno also produced several related paradoxes, including one of a flying arrow, which to go any distance, must first go half the distance, and then half of this distance, and so on, so it will never get started at all.

Everyone knows that a faster runner will always beat a slower one, given enough time to catch up, and that arrows do fly, and most discussions of these paradoxes point to their physical absurdity, but most cannot explain why they are physically absurd. The nearest that anyone has come is in seeing that it is connected with the assumption that we can divide time into the same observational units as we divide space. For example, the philosopher G. J. Whitrow writes that: 'One can, therefore, conclude that the idea of the infinite divisibility of time must be rejected, or ... one must recognize that it is... a logical fiction.'[6] Motion is 'impossible if time (and, correlatively, space) is divisible *ad infinitum*'. And the science writers Peter Coveney and Roger Highfield suggest that: 'Either one can seek to deny the notion of 'becoming', in which case time assumes essentially space-like properties; or one must reject the assumption that time, like space, is infinitely divisible into ever smaller portions.'[8] Nevertheless, there seems to be a reluctance on the part of such commentators, and others, to proceed to the logical conclusion that Zeno's paradoxes arise from the assumption that space and time are alike in their fundamental physical properties. In fact, as we have seen, it arises from the fact that, though space is indeed 'infinitely divisible into ever smaller portions', time is not divisible at all, and that the 'divisions of time' that we actually observe are only seen through the medium of space.

One of the usual strategies for tackling the problem has been to invoke calculus, and in particular to define it in terms of the 'limit' of a function as approaching a particular value. Achilles overtakes the tortoise at the point where the limit is reached. While this leads to a solution in mathematical terms, it doesn't explain the physical reason why the limit has to be invoked. As we have seen, however, there are two valid methods of calculus, only one of which involves limits. We can now see that they are, in fact, based on differentiation with respect to two different quantities with different physical properties, namely, space and time. Calculus, in principle, has nothing to do with the distinction between continuity and discontinuity, but is concerned with whether a quantity is variable or conserved. Variable quantities can be either continuous or discontinuities, and this leads to two ways of approaching the differential. If we differentiate with a discrete quantity like space, we end up with infinitesimals and non-standard analysis (cf. Chapter 2). If we differentiate with respect to time, we generate standard analysis and the theory of limits. One requires an imagined line that is absolutely continuous, the other one that is infinitely divisible. It is a classic example of the 'unreasonable effectiveness of physics in mathematics' parallelling the 'unreasonable effectiveness of mathematics in physics'. Remarkably, only the method of limits can be used to 'solve' Zeno's problem, and this is because it is really concerned with differentiation with respect to time.

Exactly the same duality *also* applies in physics, though this time it is not quite at the most primitive level, but rather at the first level of complexity. The mathematical connection between space and time does not automatically require the kind of *physical* connection supposed by Minkowski, who renamed them as one physical concept of 'space-time'. In fact the connection is, as we will show, a result of 'packaging'. This will become clearer in the next chapter. The physical identity is denied by quantum mechanics, which proclaims that time, unlike space, is not an observable. The reason will emerge only when we make quantum mechanics relativistic. In principle, the combination of space and time in a 4-vector format, while possible mathematically, cannot be done in a physical way. We are obliged to go to the nearest physical equivalent by either making time space-like or space time-like. We will argue that this is the origin of wave-particle duality, for the first solution makes everything discrete, or particle-like, while the second makes everything continuous, or wave-like. The mathematical connection diverges into two physical connections, neither of which is completely valid. Wave-particle duality would not exist if space-time was a truly physical quantity.

The duality, which runs through both classical and quantum physics, extends even to the existence of two forms of nonrelativistic quantum mechanics. Heisenberg gives us the particle-like solution, while Schrödinger gives us a version based on waves. Neither is more valid than the other, but the ideas of the theories cannot be mixed. In all cases, classical as well as quantum, the duality is absolute. Though the attempt has often been made to validate one at the expense of the other, such strategies have never succeeded. Nature gives us duality between discrete and continuous processes where space and time occur at the same level, and this is because of the fundamental difference between them due to their respective properties of discreteness and continuity.

There is just one issue to be resolved in relation to discreteness and continuity, but it is a very important one, with major implications for the way we interpret both quantum mechanics and gravity. Where does mass stand with regard to this question? There can only be one answer, either from symmetry or its intrinsic nondimensionality. Such a parameter can only be completely continuous in the same way as time. We have already stated that there is no mass which is fully discrete, even though we are accustomed to defining a rest mass or invariant mass for fundamental particles. No particle, in fact, has such a mass, for all are in motion with the relativistic energy which this involves, and even the 'rest' mass arises dynamically from the subtle quantum mechanical motion known as *zitterbewegung*. But mass-energy, in any case, is a continuum which is present at all points in space. This is seen in several different forms, in particular, the Higgs field or vacuum, consisting of 246 GeV of energy at every point in space, without which the rest masses could not be generated. It is possible, in fact, that the discrete and continuous options could be responsible for the respective ideas of the local and global, the transition from global to local gauge invariance being the point in the Higgs mechanism at which the continuous field leads to the generation of a discrete invariant mass. Besides the Higgs field, manifestations of continuous mass include the zero-point energy and even ordinary fields, which cannot be localised at points. It is the continuity of mass which is the reason for its 'unipolarity' or single sign, and for the absence of a zero or crossover point, which would indicate dimensionality.

The three dualities we have discussed in this chapter appear to be astonishingly exact, and in the long period I have been thinking about these ideas and of putting them to the test I have yet to find an exception. If they represented a truly primitive level in physics, this is exactly what we would expect to find. We would expect nature at this level to be neither totally conserved nor totally nonconserved, neither totally real nor totally imaginary, neither totally continuous nor totally discrete. In the case of the last distinction, it is impossible to imagine defining discreteness without also describing continuity. We can only know what something is if we also know what it isn't. The thing that seems to be completely excluded at this level is *extended* discreteness, except insofar as it applies to space. Continuity is often described as an 'illusion', but, at the fundamental level, where abstractions are dominant, 'illusions' or ideas are an intrinsic part of reality — we couldn't even have an idea unless it was somehow part of the abstractions which nature makes available. At this level, we would not expect to find the 'best fit' compromises that might appear at a more complex level.

While it has often been claimed that physics would fit 'reality' better if we made it totally discrete, like measurement, continuity always seems to force its way in, as in the second law of thermodynamics. The four fundamental parameters can also be interpreted in terms of system and measurement, ontology and epistemology, neither being dominant over the other. We can make the system (or theoretical superstructure) discrete, as Heisenberg did, but then continuity will appear in the measurement, as in the Heisenberg uncertainty. Alternatively, we can make the system continuous, as Schrödinger did, and find that discreteness appears in the measurement, in this case the 'collapse of the wavefunction'. As with the divisions between conserved and nonconserved, and real and imaginary parameters, this one seems to be an exact symmetry of absolute opposites.

Chapter 4

A Fundamental Symmetry

4.1 A key group

We have come a long way since we began by proposing a methodology. With only one conjecture — itself a reasonable extension from experimental evidence — we have a potential structure for the most primitive level in physics. If the structure is valid, it suggests resolutions of many of the issues connected with fundamental ideas such as space, time and 3-dimensionality, and this without a single new equation, exactly as we would expect at this level. Our next target, before we begin to explore the first level of complexity, is to find the mathematical basis, if any, behind this structure, and so tighten even further the conditions required by a fundamental theory.

We have examined three dualities between four parameters that seem to be fundamental. The pairings are between quantities that have opposite characteristics, but each is determined by the other. Duality of this kind is found throughout mathematics, and finding it in physics as well seems to suggest that the subjects are even closer in their origins than we had previously thought, and that each plays a significant role in extending the other. Remarkably, the abstract dualities seem to be responsible for characteristics that we may have once considered solidly *physical*. In the structure we have been investigating, we notice that each parameter is paired with a different partner for each duality:

Conserved **Identity**	**Nonconserved** **No identity**
Mass	Space
Charge	time

Real	**Imaginary**
Norm 1	**Norm −1**
Space	Time
Mass	Charge
Commutative	**Anticommutative**
Nondimensional	**Dimensional**
Continuous	**Discrete**
Time	Space
Mass	Charge

We see such dualities everywhere in physics and the signature of their presence is the factor 2 or $\frac{1}{2}$. Duality, as such, often seems to be more important in creating this factor than the particular duality, allowing us to switch between them. This suggests that the real explanation lies buried at a very deep level. When we have a seemingly fundamental explanation of a physical phenomenon, it may often be the case that the explanation, although correct, is not unique. A quite extraordinary example is given by the $\frac{1}{2}$ spin of the electron, and the consequent doubling of the observed magnetic moment when the electron is aligned in a magnetic field compared to the value expected from standard classical theory. Is it quantum, or relativistic, or quantum relativistic, or something else?

The answer is both none of them and all of them! There is a standard derivation from the Dirac equation using commutators. Here, the anticommutativity of the momentum operator generates a factor 2, which becomes $\frac{1}{2}$ when transferred to the other side of the equation. However, the first successful explanation of the effect was quite different, making the doubling derive from a relativistic correction of the frame of rotation, known as the Thomas precession. So, this suggests, at least, that we need either quantum theory or relativity to make the correction. Yet there is another astonishingly simple explanation of the magnetic effect using only *classical* theory. Here, we have two energy equations. One describes the kinetic energy which an object acquires during changing conditions (classically $\frac{1}{2}\,mv^2$), while the other describes the potential energy (typically, mv^2, using the virial theorem, $V = 2T$) which would maintain the system in a steady state. Using the kinetic energy equation *at the moment when the magnetic field is switched on*, we obtain the correct factor without involving either quantum theory or relativity! None of the explanations is false. All are completely true, but none uniquely so, and the real explanation is somewhere else. The fact of duality is more significant than any of the particular applications. There are a large number of phenomena in physics where this factor occurs in physics (see Chapter 10), and all of them have multiple explanations!

Explanation for spin $\frac{1}{2}$	Corresponding duality
classical kinetic energy equation	conserved/nonconserved
Thomas precession (relativity)	real/imaginary
Dirac equation (quantum mechanics)	commutative/anticommutative

Our lengthy analysis in the last chapter suggested that some symmetry was at work between the fundamental parameters. What symmetry this is can be seen by arranging the dual properties in a table:

mass	conserved	real	commutative
time	nonconserved	imaginary	commutative
charge	conserved	imaginary	anticommutative
space	nonconserved	real	anticommutative

We can make the symmetry more obvious by using the symbols, x, y and z, to represent the properties, with $-x$, $-y$ and $-z$ representing the exactly opposite properties, or, as we can conveniently describe them, the 'antiproperties':

mass	x	y	z
time	$-x$	$-y$	z
charge	x	$-y$	$-z$
space	$-x$	y	$-z$

We should immediately realise that this is one of the group structures that we examined in the second chapter, and that it represents the Klein-4 group or noncyclic group of order 4. We saw there that we could produce this group by devising a binary operation of the form:

$$x * x = -x * -x = x$$

$$x * -x = -x * x = -x$$

and similarly for y and z. Effectively, we are saying that any combination of a single property or antiproperty with itself gives the *property*; but a combination of a property with its antiproperty gives the *antiproperty*; while the combination of any property with any other property or antiproperty vanishes. This gives us the group table:

*	mass	charge	time	space
mass	mass	charge	time	space
time	charge	mass	space	time
charge	time	space	mass	charge
space	space	time	charge	mass

which, as we have said previously, has the same structure as the H_4 double algebra:

*	1	*ii*	*jj*	*kk*
1	1	*ii*	*jj*	*kk*
ii	*ii*	1	*kk*	*jj*
jj	*jj*	*kk*	1	*ii*
kk	*kk*	*jj*	*ii*	1

In the current representation, mass is the identity element, while each element is its own inverse. But this representation is not unique. We can easily rearrange the algebraic symbols to make any of space, time or charge the identity element. For example, we could have assigned the symbols in the form:

mass	$-x$	y	$-z$
time	x	$-y$	$-z$
charge	$-x$	$-y$	z
space	x	y	z

In this representation, space becomes the identity element, and the group table is now:

*	space	time	mass	charge
space	space	time	mass	charge
time	time	space	charge	mass
mass	mass	charge	space	time
charge	charge	mass	time	space

The significance of the group structure should strike the viewer immediately. If the representation is true, then we can no longer continue to do physics as though it doesn't exist. A symmetry of this kind will become an astonishingly powerful tool for generating further physical information, justifying the methodology used in deriving it. It seems to be an exact symmetry, not merely an approximation to some more fundamental truth, suggesting that we really are operating at a fundamental level. (We will soon show that the only partial conjecture that we made — concerning the nature of charge — can be fully justified by the results.) The method by which it was derived, and the principles on which it was based, suggest that it might be *exclusive*. In that case, we have an extra constraint, that there is no physical information that is not contained within it, which we can use to derive laws of physics and states of matter. Such a constraint would be even more powerful and general than those generated at present in physics by the conservation laws. Of course, because the information is at such a basic level, it will take a great deal of ingenuity to develop the more complex models which apply to physics as we mostly know it, but we hope to show that many will arise as purely natural developments.

One starting point for development is in the fact that the binary operation for the group need not be restricted to the one we have defined. If any two parameters are linked by another binary operation, then the generality of the symmetry suggests that the same binary operation must be extended to the whole group. Already, we have a binary operation between the units of space and time, and those of mass and charge, in that, if we describe them as respectively 4-vector and quaternion, their units must have a numerical relation with each other. The same numerical relation must therefore apply to all the parameters, and, because any element can be the identity and each is its own inverse, there must be a relation between the inverse units of each parameter and the units of each other, and this must apply to the units of the inverse of each parameter and itself. Ultimately, these conditions are satisfied if we can define a fundamental unit for each parameter, and, of course, these exist in the Planck length, $(\hbar G/c^3)^{1/2}$, the Planck time, $(\hbar G/c^5)^{1/2}$, the Planck mass $(\hbar c/G)^{1/2}$, and the Planck charge, $(\hbar c)^{1/2}$. The units $\hbar$, c and G, from which these are constructed, are, of course, of no fundamental significance, and merely represent the inherited choices of Babylonian astronomers and French revolutionaries.

The relations of the elements with their own inverses in creating an identity is natural given the importance of *squaring* in defining the algebra that creates them. The squared values of all the parameters have physical significance, those of space-time in creating Pythagorean or vector addition, and then combining with those of mass-charge to define physical interactions. It is typical of the nonconserved parameters of space and time that the squaring is between undefined units of the quantities, whereas that of the conserved parameters mass and charge is between specifically identifiable ones. Also, since the fundamental relations between the units have a numerical aspect, it is possible to construct new composite quantities (for example, momentum, force, energy, angular momentum), which additionally express various aspects that are characteristic of them, such as invariance, variability, dimensionality, and a real or imaginary nature.

This is necessary because the parameters only come as a package. They have no independent existence. Also, the conservation of mass and charge would have no meaning unless this came *directly linked* with variation in space and time, as the conserved quantities would be otherwise completely inaccessible. This is why we invent concepts like momentum, $\mathbf{p} = md\mathbf{r}/dt$, and extend it then to force, $\mathbf{F} = d\mathbf{p}/dt$, because we need the second order for time. The quantity is a kind of minimal level of packaging, and it becomes universal when we make it totality zero. The zeroing of fundamental composite quantities is a very general process in foundational physics. The same happens when we define force in terms of the ratio of the vector addition of squared conserved quantities to that of nonconserved ones. Again, the result is totality zero and we can relate it mathematically to the other one. We can in this relatively simple way create relationships which express many of the fundamental laws of physics, in particular those of mechanics and electromagnetic theory, in a directly mathematical form, and show that they demonstrate different aspects of the properties which the group constrains upon its elements. This important and extensive development will not be covered here in detail, since we will derive most of our results directly through quantum mechanics, but it features in *Zero to Infinity*, Chapter 8. We will, however, assume results from classical mechanics in any of its versions and from electromagnetic theory wherever needed.

The tables of properties and antiproperties and their algebraic representations also lead to some interesting reflections. First of all, we note that the parameters are completely interchangeable as abstract objects, something which is entirely within the spirit of modern algebra. This is very remarkable when we consider how they make such different impressions on our consciousness. Mass and charge appear to be tangible things, whereas we imagine space and time as more abstract. This, in fact, is an illusion, as experimental evidence seems to show that 'material objects' are ultimately composed only of abstract points. However, the illusion has had such a strong hold on the human psyche that the complete realisation that there is only abstraction is still yet to take place.

Another reflection is the strong indication here that physics at this level has a conceptual totality of exactly zero, just as we originally proposed. Of course, we don't have to use algebraic terms such as x and $-x$ to represent properties and antiproperties, and we don't have to use the expressions 'property' and 'antiproperty' for the dual concepts. However, there can be no doubt that there is a clear indication that each fundamental conceptual property in nature is negated in some sense by a property that is exactly the opposite. In general terms, there is every indication that the symmetry is absolutely exact, and that it is the exclusive source of information about the physical world.

4.2 Visual representations

If space, time, mass and charge form a group structure with an exact and unbroken symmetry, then we need to know the properties of only one of these terms to find those of all the others. These will emerge automatically from the group, like kaleidoscopic images. The arbitrary choice of which parameter becomes the 'identity' element can be seen in a number of visual representations, which, incidentally show the deep connections of the group with the fundamental nature of 3-dimensionality. There is also another analogy that we can use, in that the existence of three fundamental properties and corresponding antiproperties matches perfectly with the existence of three primary colours, red, blue and green, and three complementary secondary colours, cyan, yellow and magenta. But, again, there is no absolutely unique representation. Just as we can use any parameter as the identity element, so we can choose colours arbitrarily to represent properties and/or antiproperties (and even this designation is arbitrary).

In the colour representation, space, time, mass and charge become concentric circles, each of which is divided into three sectors. Now, for illustration, we take any primary colour to represent a property, with the corresponding secondary colour as the antiproperty. Here, we show two examples, each of which has many interpretations.

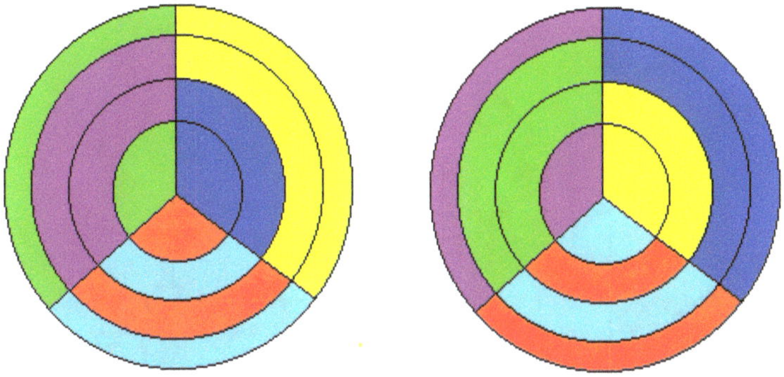

Let us, for example, take the left-hand diagram, and assume that the central triplet is mass. Then we can assume, say, that green represents real, and magenta imaginary; that blue is commutative and yellow anticommutative; and that red is conserved and cyan nonconserved. Then the next circle is imaginary, commutative and nonconserved (that is, time); the next one is imaginary, anticommutative and conserved (that is, charge); and the outer circle is real, anticommutative and nonconserved (that is, space). Each sector, in any version, always adds to zero, represented by white. The representation is striking for showing that, in many ways, it is the pattern that is important, rather than the specifics. The structure only makes sense as a complete package.

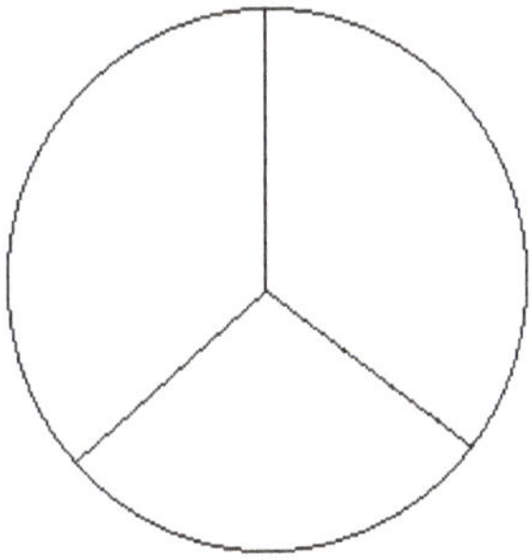

As an alternative to using colour, we could make direct use of the labels x, y, z to represent axes in 3-dimensional space. This time, it is the

+ and − directions that represent property and antiproperty. The four parameters then become equivalent to lines drawn from the centre of a cube to four of its corners.

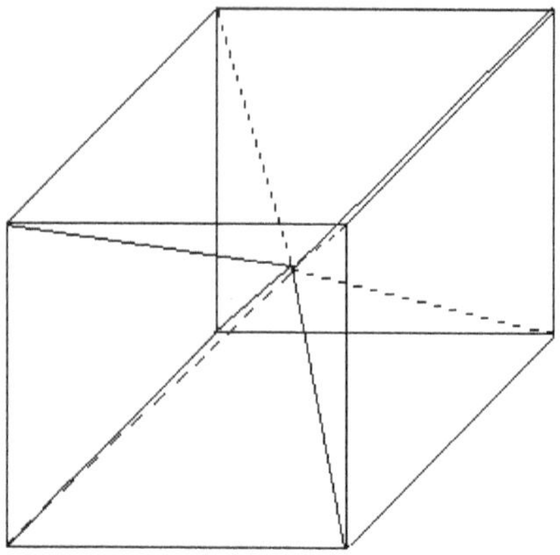

The dotted lines then represent the alternative arrangement of the signs of x, y and z (say, by switching the signs of any of x, y or z) which we showed in the second chapter would lead to the same group:

$$\begin{array}{rrr} -x & -y & -z \\ x & y & -z \\ -x & y & z \\ x & -y & z \end{array}$$

This dual arrangement (also seen in the second of our colour examples) might also be physical in some sense, though not primary. A good candidate for this occurs in the relativistic quantum mechanics of the Dirac equation, where there is an apparent reversal in some of the characteristics of the fundamental parameters. The easiest to change is the real/imaginary distinction. In this case, we have:

mass†	$-x$	$-y$	$-z$
time†	x	y	$-z$
charge†	$-x$	y	z
space†	x	$-y$	z

or

mass†	conserved	imaginary	commutative
time†	nonconserved	real	commutative
charge†	conserved	real	anticommutative
space†	nonconserved	imaginary	anticommutative

The reason for this will become clear when we discuss the packaging of the parameter group.

In yet another representation, the parameters can be placed at the vertices of a regular tetrahedron. The six edges in the primary and secondary colours (R, G, B and M, C, Y, in the diagram) now represent the respective properties and antiproperties, or vice versa. Alternatively, since the tetrahedron is a dual structure in itself, we can represent the parameters by its *faces*. Once again, an extra duality appears, as with the colour and cubic representations.

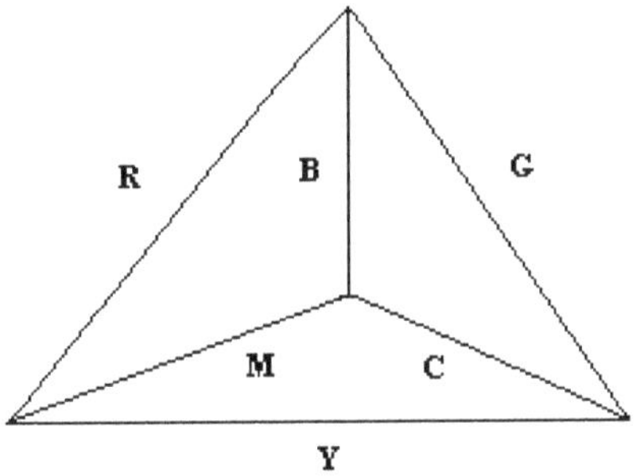

In each case, there is something like a C_2 symmetry between the dual D_2 structures, and the $C_2 \times D_2$ of order 8 creates a larger structure of the form:

*	M	C	S	T	$M^\dagger$	$C^\dagger$	$S^\dagger$	$T^\dagger$
M	M	C	S	T	$M^\dagger$	$C^\dagger$	$S^\dagger$	$T^\dagger$
C	C	$M^\dagger$	T	$S^\dagger$	$C^\dagger$	M	$T^\dagger$	S
S	S	$T^\dagger$	$M^\dagger$	C	$S^\dagger$	T	M	$C^\dagger$
T	T	S	$C^\dagger$	$M^\dagger$	$T^\dagger$	$S^\dagger$	C	M
$M^\dagger$	$M^\dagger$	$C^\dagger$	$S^\dagger$	$T^\dagger$	M	C	S	T
$C^\dagger$	$C^\dagger$	M	$T^\dagger$	S	C	$M^\dagger$	T	$S^\dagger$
$S^\dagger$	$S^\dagger$	T	M	$C^\dagger$	S	$T^\dagger$	$M^\dagger$	C
$T^\dagger$	$T^\dagger$	$S^\dagger$	C	M	T	S	$C^\dagger$	$M^\dagger$

Remarkably, this structure is identical to that of the quaternion group (Q):

$*$	1	i	j	k	-1	$-i$	$-j$	$-k$
1	1	i	j	k	-1	$-i$	$-j$	$-k$
i	i	-1	k	$-j$	$-i$	1	$-k$	j
j	j	$-k$	-1	i	$-j$	k	1	$-i$
k	k	j	$-i$	-1	$-k$	$-j$	i	1
-1	-1	$-i$	$-j$	$-k$	1	i	j	k
$-i$	$-i$	1	$-k$	j	i	-1	k	$-j$
$-j$	$-j$	k	1	$-i$	j	$-k$	-1	i
$-k$	$-k$	$-j$	i	1	k	j	$-i$	-1

4.3 Two spaces?

In relation to the 'unreasonable effectiveness' of mathematics in physics and of physics in mathematics, it is significant that what we think of as *physical* properties — and the only ones that actually exist at the fundamental level — can be expressed almost entirely as pure algebra. Two of the dualities — real and imaginary, and commutative and noncommutative — are obviously so. The third — conserved and nonconserved — is almost certainly in the same category. It is probable that the algebraic reason behind it is something like the fact that the nonconserved parameters — time and space — are constructed partly from an incomplete quaternion group, i.e. complex numbers, based on the imaginary 'pseudoscalar' i. This aspect of complex numbers is more apparent when the algebra is constructed from a universal rewrite system algebra, as was done in work carried out with my colleague Bernard Diaz.[7] A similar kind of reasoning concerning incompletion was the idea which initially led to the discovery of quaternions.

The four fundamental parameters encompass four separate algebraic systems. These algebras also automatically generate subalgebras, which include the real number algebra for scalar magnitudes of all quantities; and, for space, the pseudovector ($\equiv$ quaternion) algebra for area, and pseudoscalar ($\equiv$ complex) algebra for volume.

		algebra	*subalgebras*		
Mass	Real numbers	1			
Time	Imaginary numbers	i	1		
Charge	Quaternions	$\boldsymbol{i}, \boldsymbol{j}, \boldsymbol{k}$	1		
Space	Vectors	$\mathbf{i}, \mathbf{j}, \mathbf{k}$	1	i	$i\mathbf{i}, i\mathbf{j}, i\mathbf{k}$

We can, in fact, restructure these algebras and subalgebras using the vector/bivector/trivector terminology of Clifford algebra:

Space	i, j, k	vector		
	ii, ij, ik $\equiv$ i, j, k	bivector	pseudovector	quaternion
	i	trivector	pseudoscalar	
	1	scalar		
Charge	i, j, k $\equiv$ ii, ij, ik	bivector	pseudovector	quaternion
	1	scalar		
Time	i	trivector	pseudoscalar	
	1	scalar		
Mass	1	scalar		

What is immediately noticeable is that the algebras of mass, time and charge are, mathematically, subalgebras of the algebra of space. Now, if we put together the three subalgebras, we obtain an equivalent of the algebra itself. The combination of i and ii, ij, ik (with or without 1) will create the missing i, j, k. In effect, putting charge, time, and mass together gives us a mathematical equivalent to another space (i, j, k) alongside the real space (i, j, k). Physically, this composite is not a space, because it is not a single quantity, and so it will never be measurable or observable in the same way as space. However, mathematically it is the same.

We can consider it as an *antispace*. It incorporates all the things that are *not* space, allowing us to equate the totality of space and this composite object to zero. Using the convention that space is the identity element, the respective algebraic sums for the properties would be: space (x, y, z) and antispace ($-x$, $-y$, $-z$). The same could, of course, be done for time, mass or charge, but as these are not measurable quantities, the effect is rather less significant. Overall, we see that the symmetry between the parameters seems to be telling us that nature or the universe is, as we have suspected from the beginning, a conceptual 'nothing' or zero, with no defining characteristics. It isn't even possible to decide whether we should view it ontologically (the 'God's eye' view) or epistemologically (the view of an observer). And the factor 2 that appears everywhere in physics comes from the fact that any defined aspect of nature also produces a mirror image of itself which negates its existence.

We have already noted the increasing tendency in physics to suggest that the universe has some kind of zero sum of quantities like force and energy, but it seems to me that the various suggestions stop short of the possibility that there is absolutely *nothing at all* in the 'universe' or in 'nature', even *conceptually*. Of course, this may at first seem startling, because we seem to be surrounded by 'something' everywhere we look, but really, looking from the *inside*, we could have no real idea about 'nothing' referring to nature as a whole. We can write down equations with zero on the right hand side, but we can't realise any of them physically. Getting 'close to zero', if it isn't actually zero, isn't really close at all. Clearly, the zeroing is much more subtly arranged than we would find simply by taking something like $1 - 1 = 0$, as we can see from the group symmetry that we have uncovered, that suggests zero without immediate cancellation. But, however we perceive it, totality zero is an especially powerful constraint because it forces us into a holistic view of the universe, such as quantum mechanics seems to require.

4.4 A unified algebra

To take physics further we need to put together the 'package' which incorporates all the individual components into a coherent unified system. This creates the first level of complexity. We have already outlined the mathematics necessary to do this. We can, for example, take the Clifford algebra approach, and put together two vector spaces, which are commutative with each other, with fundamental units consisting of + and – versions of

i	**j**	**k**	*i***i**	*i***j**	*i***k**	*i*	1
i	**j**	**k**	*i***i**	*i***j**	*i***k**	*i*	1
vector			*bivector*			*trivector*	*scalar*

The product of each term with every other, or tensor product, consists of 64 terms, which are + and – values of the following:

i	**j**	**k**	*i***i**	*i***j**	*i***k**	*i*	1
i	**j**	**k**	*i***i**	*i***j**	*i***k**		
ii	**ij**	**ik**	*i***ii**	*i***ij**	*i***ik**		
ji	**jj**	**jk**	*i***ji**	*i***jj**	*i***jk**		
ki	**kj**	**kk**	*i***ki**	*i***kj**	*i***kk**		

We could equally well have begun with the four algebras of space, time, mass and charge:

i j k	*i*	1	*i j k*
space	time	mass	charge
vector	*pseudoscalar*	*scalar*	*quaternion*

This would give us the completely equivalent vector-quaternion algebra, which would emerge from exchanging *ii*, *ij*, *ik* for i, j, k and *i*, *j*, *k* for *i*i, *i*j, *i*k, and which requires + and – values of:

i	j	k	*i*i	*i*j	*i*k	*i*	1
i	*j*	*k*	*ii*	*ij*	*ik*		
i*i*	i*j*	i*k*	*i*i*i*	*i*i*j*	*i*i*k*		
j*i*	j*j*	j*k*	*i*j*i*	*i*j*j*	*i*j*k*		
k*i*	k*j*	k*k*	*i*k*i*	*i*k*j*	*i*k*k*		

We have already obtained these algebras and identified them as a group of order 64. Here, we have 8 generators of the algebra, which, using the two vector spaces, i, j, k, and i, j, k, we could reduce to 6. But neither of these is the minimum, which we have already shown reduces to 5, all of course, elements of the group. This can be done in many ways, but all those that incorporate all the base elements look something like

ik *i*i *i*j *i*k *j*

or

*i*k ii ij ik j

All the sets of 5 generators have the same pattern, as we have seen by splitting up the 64 units into 1, –1, *i* and –*i*, and 12 sets of 5 generators, each of which generates the entire group:

1	*i*				–1	–*i*			
*i*i	*i*j	*i*k	*ik*	*j*	–*i*i	–*i*j	–*i*k	–*ik*	–*j*
*j*i	*j*j	*j*k	*ii*	*k*	–*j*i	–*j*j	–*j*k	–*ii*	–*k*
*k*i	*k*j	*k*k	*ij*	*i*	–*k*i	–*k*j	–*k*k	–*ij*	–*i*
*i*i*i*	*i*i*j*	*i*i*k*	*i*k	j	–*i*i*i*	–*i*i*j*	–*i*i*k*	–*i*k	–j
*i*j*i*	*i*j*j*	*i*j*k*	*i*i	k	–*i*j*i*	–*i*j*j*	–*i*j*k*	–*i*i	–k
*i*k*i*	*i*k*j*	*i*k*k*	*i*j	i	–*i*k*i*	–*i*k*j*	–*i*k*k*	–*i*j	–i

Even this arrangement is not unique, but any rearrangement would retain the same pattern in which the symmetry of one of the two 3-dimensional structures (i, j, k or *i*, *j*, *k*; and i, j, k) is broken while the symmetry of the other is preserved. Physics always tends to go for the most minimal representation, and though something like

ik *i*i *i*j *i*k *j*

does not appear to be as symmetrical at first sight as

$$\mathrm{i} \quad \mathrm{j} \quad \mathrm{k} \qquad i \qquad 1 \qquad i \quad j \quad k$$

it contains the same information, and, ultimately, the same symmetries. It is thus in creating the minimum packaging for the information contained in the parameter group that we find the ultimate explanation of why the symmetry of charge is broken, at the first level of complexity (packaging), whereas that of space is not. The process is completely dual, so it would be quite possible to create a physics in which the process was reversed, and the geometry of space was altered rather than the charge structure, say using the structure of a Finsler geometry, but, for comparison with the bulk of physics as we know it, it seems more convenient to retain the symmetry of space rather than that of charge.

The symmetry between the three components of charge and their interactions can be seen to be broken at the level of observation, that is, when we package it with space. To preserve the symmetry of the observed quantity, real space (that of i, j, k), we necessarily have to break the symmetry of 'charge' (i, j, k) or the unobservable mathematical 'space' (i, j, k) that links charge with mass and time. Effectively, starting with the 8 units needed for the 4 parameters:

$$\begin{array}{cccc} i & \mathrm{i} \quad \mathrm{j} \quad \mathrm{k} & 1 & i \quad j \quad k \\ \text{time} & \text{space} & \text{mass} & \text{charge} \end{array}$$

we 'compactify' to the 5 generators by removing the three 'charge' units and attaching one to each of the other three parameters:

$$\begin{array}{ccccc} i & \mathrm{i} & \mathrm{j} & \mathrm{k} & 1 \\ k & & i & & j \end{array}$$

As a result, we create 3 new 'composite' parameters, each of which has aspects of time, space or mass, but also some characteristics of charge.

$$\begin{array}{ccccc} ik & i\mathrm{i} & i\mathrm{j} & i\mathrm{k} & j \\ E & p_x & p_y & p_z & m \end{array}$$

We can attach to these unit structures any *scalar* labels we like, and here we select those that will subsequently be identified as those for energy, momentum and rest mass. The significant thing here is that these quantities are defined by their algebraic units not by their scalar values. Since we started only with space, time, mass and charge, this becomes the *first appearance* of these *conjugate* quantities in physics, and it would seem that the superposition of two sets of parameters with different characters to create generators for the group combining their algebras actually *creates* them. It also simultaneously fixes quantization and relativity as fundamental components of the package, each of these being effectively the establishment of numerical relations between the units of previously unrelated physical quantities.

The new structure we have created is essentially the one that we normally describe as phase space, but it is not independent of either of the 'spaces' that go into its making. This means that the quantities in the conjugate pairings, time and energy, and space and momentum, are not actually independent, for the set involving energy and momentum is in part created from the more primitive set involving time and space. While fully independent quantities are commutative with each other, dependent quantities are not. Energy and time are therefore anticommutative at the level of the most fundamental units, as are momentum and space. This is exactly what is expressed in Heisenberg's uncertainty principle: $2 \times$ the product of the fundamental units of the two anticommutative terms produces the most fundamental quantum unit of their combination, $\hbar = h/2\pi$, the quantum unit of angular momentum.

4.5 Nilpotency

Physics operates in such a way that the total package of all information is zero, and the combined structure we have created by packaging the entire source of information available to us, $(\boldsymbol{ik}E + \boldsymbol{i}\mathbf{i}p_x + \boldsymbol{i}\mathbf{j}p_y + \boldsymbol{i}\mathbf{k}p_z + \boldsymbol{j}m)$, becomes a norm 0 object, or a nilpotent. So

$$(\boldsymbol{ik}E + \boldsymbol{i}\mathbf{i}p_x + \boldsymbol{i}\mathbf{j}p_y + \boldsymbol{i}\mathbf{k}p_z + \boldsymbol{j}m)^2 = E^2 - p^2 - m^2 = 0$$

which immediately creates the numerical relations we require between all the parameters. Now, from a fundamental point of view, we can begin to see that the nilpotent structure is equivalent to creating a point source or charge singularity in 3-dimensional space. In effect, we combine two spaces, or space and antispace, to effectively cancel and create a region in which the 'spatial' extent is zero. Through the nilpotent condition, the two spaces share dual information, though it is differently organized in each. The observed 3-dimensional space becomes multiply-connected because it is acting as *two* spaces, and only one of which is observed. The space that remains unobserved is described as 'vacuum space' in quantum mechanics. A circuit of a closed path in real space will require a double rotation to return to the starting point because it is only in this space for half the time. The charge singularity will itself be a multiply-connected space, and require a double circuit, which will manifest itself as spin $\frac{1}{2}$.

We can regard the 5 group generators as the most efficient packaging of all the information contained in the group structure of space, time, mass and charge, and codified in their algebraic structures. We should be able to use it to generate the physics that we know is contained in the interactions between fermions, in particular the Dirac equation and the relativistic quantum mechanics of fermions and bosons. In fact, this emerges in an extraordinarily transparent form, in which many developments follow immediately from the algebraic structure. This will be covered mostly in the following chapter, but it will be useful to do a preliminary analysis here. The apparently classical expression

$$(\boldsymbol{ik}E + \boldsymbol{i}\mathbf{i}p_x + \boldsymbol{i}\mathbf{j}p_y + \boldsymbol{i}\mathbf{k}p_z + \boldsymbol{j}m)(\boldsymbol{ik}E + \boldsymbol{i}\mathbf{i}p_x + \boldsymbol{i}\mathbf{j}p_y + \boldsymbol{i}\mathbf{k}p_z + \boldsymbol{j}m) = 0$$

or

$$(\boldsymbol{ik}E + \boldsymbol{i}\mathbf{p} + \boldsymbol{j}m)(\boldsymbol{ik}E + \boldsymbol{i}\mathbf{p} + \boldsymbol{j}m) = 0$$

can be immediately restructured as relativistic quantum mechanics using a canonical quantization of the first bracket ($E \to i\partial/\partial t$, $\mathbf{p} \to -i\nabla$) and its application to a phase factor, which, for a free particle, would be $e^{-i(Et-\mathbf{p}\cdot\mathbf{r})}$. So that

$$(-\boldsymbol{ik}\partial/\partial t - \boldsymbol{i}\nabla + \boldsymbol{j}m)(\boldsymbol{ik}E + \boldsymbol{i}\mathbf{p} + \boldsymbol{j}m)e^{-i(Et-\mathbf{p}\cdot\mathbf{r})} = 0$$

which, as we will demonstrate in the next chapter, is the Dirac equation for the free fermion. In effect, the equation shows the simultaneous application of the dual 'spaces' involved in the nilpotent structure, the 'amplitude' term $(\boldsymbol{ik}E + \boldsymbol{i}\mathbf{p} + \boldsymbol{j}m)$ representing the localised, real space of the point particle, and the operator $(-\boldsymbol{ik}\partial/\partial t - \boldsymbol{i}\nabla + \boldsymbol{j}m)$ acting on the phase factor $e^{-i(Et-\mathbf{p}\cdot\mathbf{r})}$ representing the variation over the delocalised vacuum space. The phase factor in this form gives us the expression for a space and time that can be varied without restriction, and the operator acting on it sets up the conservation conditions that have to be applied simultaneously. Ultimately, we will see that the amplitude and phase are not independent information. The entire information is incorporated within the operator $(-\boldsymbol{ik}\partial/\partial t - \boldsymbol{i}\nabla + \boldsymbol{j}m)$, which sets up the 'nonconservation' conditions for space and time, which lead to the creation of the energy and momentum as conserved quantities, in addition to angular momentum (for we will eventually recognise that the term $(i\boldsymbol{k}E + \boldsymbol{i}\mathbf{p} + \boldsymbol{j}m)$ is, in principle, an angular momentum operator).

4.6 The symmetry-breaking between charges

The packaging process affects time, space and mass by creating the energy–momentum–rest mass conjugate. But it must also affect charge, for it simultaneously creates three new 'charge' units, which take on the respective characteristics of the parameters with which they are associated.

$i\boldsymbol{k}$	$\boldsymbol{i}\mathbf{i}$ $\boldsymbol{i}\mathbf{j}$ $\boldsymbol{i}\mathbf{k}$	$\boldsymbol{j}$
weak charge	strong charge	electric charge
pseudoscalar	*vector*	*scalar*

In the Standard Model, the symmetry between the weak, strong and electric interactions is broken in such a way that they respond respectively to the symmetry groups $SU(2)$, $SU(3)$ and $U(1)$. These group structures, though well established on the basis of a large amount of experimental work, have no fundamental explanation in the Standard Model. However, it should now be possible to see that they are generated through the 2-component pseudoscalar ($SU(2)$), 3-component vector ($SU(3)$) and single component scalar ($U(1)$) nature of the weak, strong and electric charges as they are incorporated within the nilpotent structure. This will become much more explicit after we have established a system of quantum mechanics.

When we extend the analysis to quantum theory, we will see that the modification of charge shows the continuous, nonlocal or vacuum side of the compactification process, while the compactification of time, space and mass to energy, momentum and rest mass shows the discrete or local. Local and nonlocal, however, are not separate things. Neither is defined without the other. Local interactions can be seen to have nonlocal consequences, while nonlocal interactions have local consequences.

4.7 The parameters in the dual group

We can now return to some specific issues which we have so far left unresolved. One is the nature of the dual group to space, time, mass and charge. The extra quaternion units in the expression $(i\boldsymbol{k}E + \boldsymbol{i}\mathbf{p} + \boldsymbol{j}m)$ clearly change the norm of the time-like term $(i\boldsymbol{k}E)$ from –1 to 1, and those of the space-like and mass-like terms ($\boldsymbol{i}\mathbf{p}$ and $\boldsymbol{j}m$) from 1 to –1, so making the quantized energy and momentum and rest mass terms equivalent to time†, space† and mass†. The same would be true if we used the nilpotent structure $(ikt + i\mathbf{r} + \boldsymbol{j}\tau)$ for the relativistic space-time invariance, where τ is the proper time. The quantized angular momentum would then be equivalent to the charge† term, in line with the already established link between charge and angular momentum. The group of order 8 incorporating the D_2 parameter group and its mathematical dual, which is isomorphic to the quaternions, would then be the quantized phase space for the fermion.

Now, if mass, charge, time and space form a group of order 4, then the group of their base units (1, i, j, k, i, i, j, k), or (m, s, e, w, t, x, y, z), expressed in the form of complexified double quaternions, rather than the more usual quaternion-4-vector combination (1, $\boldsymbol{i}$, $\boldsymbol{j}$, $\boldsymbol{k}$, i, $\mathbf{i}$, $\mathbf{j}$, $\mathbf{k}$) could be said to be that of a 'broken octonion'. (It is also intriguingly close to Penrose's twistor structure in having 4 'real' parts (norm 1) and 4 'imaginary' parts (norm –1), though the additional structure here turns out to be crucial in separating out two sets of 3-dimensional objects and two full vector spaces.) The breaking doesn't occur in the sense that a large structure which is fundamental is exposed to a symmetry-breaking 'mechanism', but because the large structure is made up out of units with an independent origin, which have asymmetric aspects from the beginning. Symmetry-breaking seems to come from the bottom up, not from the top down. In fact, the complexified double quaternion structure readily maps onto that of the octonions (in the second following table), with the antiassociative multiplications excluded from the physical meaning which is created within the separate parts from which the structure was made. Since the octonion structure is the basis of some of the higher groups such as E_8 which are thought to be significant in generating the particle spectrum, it is relevant that the brokenness which has to be introduced into such theories would here be carried forward to the higher groups from the most basic level.

*	*m*	*s*	*e*	*w*	*t*	*x*	*y*	*z*
m	*m*	*s*	*e*	*w*	*t*	*x*	*y*	*z*
s	*s*	–*m*	*w*	–*e*	*x*	–*t*	–*z*	*y*
e	*e*	–*w*	–*m*	*s*	*y*	*z*	–*t*	–*x*
w	*w*	*e*	–*s*	–*m*	*z*	–*y*	*x*	–*t*
t	*t*	–*x*	–*y*	–*z*	–*m*	*s*	*e*	*w*
x	*x*	*t*	–*z*	*y*	–*s*	–*m*	–*w*	*e*
y	*y*	*z*	*t*	–*x*	–*e*	*w*	–*m*	–*s*
z	*z*	–*y*	*x*	*t*	–*w*	–*e*	*s*	–*m*

*	1	i	j	k	e	f	g	h
1	1	*i*	*j*	*k*	*e*	*f*	*g*	*h*
i	*i*	–1	*k*	–*j*	*f*	–*e*	–*h*	*g*
j	*j*	–*k*	–1	*i*	*g*	*h*	–*e*	–*f*
k	*k*	*j*	–*i*	–1	*h*	–*g*	*f*	–*e*
e	*e*	–*f*	–*g*	–*h*	–1	*i*	*j*	*k*
f	*f*	*e*	–*h*	*g*	–*i*	–1	–*k*	*j*
g	*g*	*h*	*e*	–*f*	–*j*	*k*	–1	–*i*
h	*h*	–*g*	*f*	*e*	–*k*	–*j*	*i*	–1

4.8 Conservation of angular momentum and conservation of type of charge

The other issue is particularly significant because it illustrates the predictive value of the fundamental methodology. Earlier, we predicted a quite extraordinary result as a consequence of Noether's theorem. This equated the conservation of angular momentum or the rotational symmetry of space with the conservation of *type* of charge, i.e. the inability of weak, strong and electric charges to transform into each other. The result looks impossible to demonstrate or to fit to a mathematical description, but now we can give the explanation. Essentially, angular momentum conservation is made up of *three separate conservation laws* which are completely independent but all required at the same time. For angular momentum to be conserved, we have to separately conserve the magnitude, the direction, and the handedness (i.e. whether the rotation is right- or left-handed), and the symmetries we require for these conservation laws are the $U(1)$, $SU(3)$ and $SU(2)$ symmetries involved with the electric, strong and weak charges. In principle, these symmetries are versions of the spherical symmetry of 3-dimensional space around a point charge. Spherical symmetry, they say, is preserved by a rotating system

	whatever the length of the radius vector	$U(1)$;
	whatever system of axes we choose	$SU(3)$;
and	whether we choose to rotate the system left- or right-handed	$SU(2)$.

Conservation of charge is thus the same thing as the conservation of spherical symmetry for a point source, and it has to preserve all three aspects. As we have seen from our analysis of symmetry-breaking, the $SU(3)$ and $SU(2)$ aspects are dealt with by the respective strong and weak charges, with their vector and pseudoscalar characteristics. These are additional to the $U(1)$ symmetry, to which all three charges contribute (just as they do to the Coulomb interaction) because all three charges also have scalar characteristics. The electric charge is unique, however, in contributing only to this symmetry. So all three charges have to be conserved independently of each other, in the same way as the direction, handedness and magnitude of the angular momentum. It must be one of the strongest possible tests of a theory to predict such a totally unexpected result and then to find a simple reason to why it must be valid.

Chapter 5

Nilpotent Quantum Mechanics I

5.1 The Dirac equation

Many laws of physics can be seen as purely abstract statements about relations between fundamental quantities and the properties that emerge from the parameter group. While it is possible to derive a number of them by individual arguments, a more significant development is to try to explain the most fundamental virtually at one go from the packaging process that generates the algebraic group of order 64 connecting two vector spaces. This group turns out to be very significant indeed as it lies behind what seems to be the most fundamental equation in physics: Dirac's relativistic quantum mechanical equation for the fermion. At its deepest level, physics is only about fermions and their interactions, bosons, the only other particles, being generated by the interactions, and an equation for the fermion in its most general form will necessarily contain, explicitly or implicitly, many of the other principles that are significant in physics.

We have already made the claim that both the Dirac equation and the structure of the fermion can be derived, in principle, from the most efficient packaging of the parameters of space, time, mass and charge. We now need to justify this claim by testing it against the Dirac equation as normally presented. Dirac started with the Schrödinger equation in which a second-order differential with respect to space is linked with a first-order differential with respect to time. Clearly, this is not relativistic, which requires both terms to be of the same order. Dirac needed a quantum version of Einstein's relativistic energy–momentum equation, $E^2 - p^2 - m^2 = 0$, but quantizing this in the usual way gave the Klein–Gordon equation, with second-order differentials for space and time, which didn't give the required behaviour for the fermion. He realised what was needed was an equation that was *first-order* in the two variables. Effectively, this meant 'square-rooting' a second-order equation to give *the same number* of linear terms. Now, this can be done if you use anticommuting operators, which will eliminate the cross terms when the square root is squared again. The choice that may seem obvious now — quaternions — was not available to Dirac because of the general prohibition on their use, though he did have Pauli matrices. Dirac chose a set of 4×4 matrices and found an equation which worked spectacularly, especially in its automatic inclusion of the fermionic property of half-integral spin. In one of its main forms it looks something like:

$$(\gamma^\mu \partial_\mu + im)\psi = \left(\gamma^0 \frac{\partial}{\partial t} + \gamma^1 \frac{\partial}{\partial x} + \gamma^2 \frac{\partial}{\partial y} + \gamma^3 \frac{\partial}{\partial z} + im\right)\psi = 0$$

The γ terms are the matrices, and they are mutually anticommutative; γ^0 is a square root of 1 (actually the identity matrix I), and γ^1, γ^2, γ^3 are square roots of -1 (actually of $-I$), which are mutually orthogonal components of an object with vector properties ($\boldsymbol{\gamma}$) (the origin of the automatic inclusion of fermionic spin). The algebra is closed, but only with the addition of another matrix, γ^5, which is another square root of I, and is anticommutative to all the other matrices, but which does not appear in the equation. From the five generators, γ^0, γ^1, γ^2, γ^3, γ^5, we create a group of 64 possible combinations. The composition of the matrices is not unique, but it is usual to construct them from 2×2 Pauli matrices in the form

$$\boldsymbol{\gamma} = \begin{pmatrix} 0 & \boldsymbol{\sigma} \\ -\boldsymbol{\sigma} & 0 \end{pmatrix}, \quad \gamma_0 = \begin{pmatrix} I & 0 \\ 0 & -I \end{pmatrix}, \quad \gamma_5 = \begin{pmatrix} 0 & -I \\ -I & 0 \end{pmatrix},$$

$$\gamma_1 = \begin{pmatrix} 0 & \sigma_x \\ -\sigma_x & 0 \end{pmatrix}, \quad \gamma_2 = \begin{pmatrix} 0 & \sigma_y \\ -\sigma_y & 0 \end{pmatrix}, \quad \gamma_3 = \begin{pmatrix} 0 & \sigma_z \\ -\sigma_z & 0 \end{pmatrix}$$

which can be expanded to:

$$\gamma_1 = \begin{pmatrix} 0 & 0 & 0 & 1 \\ 0 & 0 & 1 & 0 \\ 0 & -1 & 0 & 0 \\ -1 & 0 & 0 & 0 \end{pmatrix}, \quad \gamma_2 = \begin{pmatrix} 0 & 0 & 0 & -i \\ 0 & 0 & i & 0 \\ 0 & i & 0 & 0 \\ -i & 0 & 0 & 0 \end{pmatrix},$$

$$\gamma_3 = \begin{pmatrix} 0 & 0 & 1 & 0 \\ 0 & 0 & 0 & -1 \\ -1 & 0 & 0 & 0 \\ 0 & 1 & 0 & 0 \end{pmatrix}, \quad \gamma_0 = \begin{pmatrix} 1 & 0 & 0 & 0 \\ 0 & 1 & 0 & 0 \\ 0 & 0 & -1 & 0 \\ 0 & 0 & 0 & -1 \end{pmatrix},$$

$$I = \begin{pmatrix} 1 & 0 & 0 & 0 \\ 0 & 1 & 0 & 0 \\ 0 & 0 & 1 & 0 \\ 0 & 0 & 0 & 1 \end{pmatrix}$$

The use of 4×4 matrices meant that the wavefunction ψ had to be expanded to a column vector with four components, which then meant that the equation had 4 'solutions', which could be identified with (fermion and antifermion) $\times$ (spin up and spin down). In fact it has the form of a spinor or a single component wavefunction multiplied by a spinor. If we make an explicit use of all the relevant matrices in the differential operator, the equation now becomes:

$$\begin{pmatrix} \frac{\partial}{\partial t} + im & 0 & \frac{\partial}{\partial z} & \frac{\partial}{\partial x} - i\frac{\partial}{\partial y} \\ 0 & \frac{\partial}{\partial t} + im & \frac{\partial}{\partial x} + i\frac{\partial}{\partial y} & -\frac{\partial}{\partial z} \\ -\frac{\partial}{\partial z} & -\frac{\partial}{\partial x} + i\frac{\partial}{\partial y} & -\frac{\partial}{\partial t} + im & 0 \\ -\frac{\partial}{\partial x} - i\frac{\partial}{\partial y} & \frac{\partial}{\partial z} & 0 & -\frac{\partial}{\partial t} + im \end{pmatrix} \begin{pmatrix} \psi_1 \\ \psi_2 \\ \psi_3 \\ \psi_4 \end{pmatrix} = 0 \quad (1)$$

Now, Dirac seems nowhere to have investigated the true origin of his matrices, which, significantly form a group of order 64, including the identity matrix I and both original and complexified versions of each matrix (generically, M and iM). It is not a thing that you find in textbooks either, but, in fact, the way to construct the complete set of 4×4 matrices, *in all their variant forms*, is by using *two* sets of Pauli matrices, $\boldsymbol{\sigma}^1$, $\boldsymbol{\sigma}^2$, $\boldsymbol{\sigma}^3$ and $\boldsymbol{\Sigma}^1$, $\boldsymbol{\Sigma}^2$, $\boldsymbol{\Sigma}^3$, which are commutative with each other. Again, this forms a group of order 64, with a minimum of 5 generating elements, which may be something like

$$\boldsymbol{\Sigma}^1\boldsymbol{I},\ i\boldsymbol{\sigma}^1\boldsymbol{\Sigma}^3,\ i\boldsymbol{\sigma}^2\boldsymbol{\Sigma}^3, i\boldsymbol{\sigma}^3\boldsymbol{\Sigma}^3,\ i\boldsymbol{\Sigma}^2\boldsymbol{I}$$

or

$$\boldsymbol{\Sigma}^3 \boldsymbol{I},\ i\boldsymbol{\sigma}^1\boldsymbol{\Sigma}^1,\ i\boldsymbol{\sigma}^2\boldsymbol{\Sigma}^1,\ i\boldsymbol{\sigma}^3\boldsymbol{\Sigma}^1,\ i\boldsymbol{\Sigma}^2\boldsymbol{I}$$

and which, when we multiply them out, will be found to be identical in structure to γ^0, γ^1, γ^2, γ^3, γ^5. Now we know that our two Pauli algebras, $\boldsymbol{\sigma}^1$, $\boldsymbol{\sigma}^2$, $\boldsymbol{\sigma}^3$ and $\boldsymbol{\Sigma}^1$, $\boldsymbol{\Sigma}^2$, $\boldsymbol{\Sigma}^3$, are exactly isomorphic to the two sets of vector units, i, j, k, and **i**, **j**, **k**, or the two sets of complexified quaternions that we have used before. Even using the two Pauli algebras, we can see that in defining the group generators, we preserve the symmetry of one algebra ($\boldsymbol{\sigma}$) and break that of the other ($\boldsymbol{\Sigma}$).

5.2 The nilpotent Dirac equation

The standard form of the Dirac equation is not completely symmetric, as γ^5 is excluded although it is needed as one of the generators of the algebra. We can remedy this simply by premultiplying by $-i\gamma^5$:

$$-i\gamma^5(\gamma^\mu\partial_\mu + im)\,\psi = -i\gamma^5\left(\gamma^0\frac{\partial}{\partial t} + \gamma^1\frac{\partial}{\partial x} + \gamma^2\frac{\partial}{\partial y} + \gamma^3\frac{\partial}{\partial y} + im\right)\psi = 0.$$

Using the double Pauli version of the γ matrices we can convert this to

$$\boldsymbol{\Sigma}^2 I\left(\boldsymbol{\Sigma}^1 I\frac{\partial}{\partial t} + i\boldsymbol{\Sigma}^3\boldsymbol{\sigma}^1\frac{\partial}{\partial x} + i\boldsymbol{\Sigma}^3\boldsymbol{\sigma}^2\frac{\partial}{\partial y} + i\boldsymbol{\Sigma}^3\boldsymbol{\sigma}^3\frac{\partial}{\partial y} + im\right)\psi = 0$$

which then becomes

$$\left(-i\boldsymbol{\Sigma}^3 I\frac{\partial}{\partial t} - \boldsymbol{\Sigma}^1\boldsymbol{\sigma}^1\frac{\partial}{\partial x} - \boldsymbol{\Sigma}^1\boldsymbol{\sigma}^2\frac{\partial}{\partial y} - \boldsymbol{\Sigma}^1\boldsymbol{\sigma}^3\frac{\partial}{\partial y} + i\boldsymbol{\Sigma}^2 Im\right)\psi = 0,$$

and we can now begin to appreciate the new degree of symmetry we have created. Not only are all 5 generators included in the equation and one assigned to each term of the operator, but also the units of the two vector spaces, $\boldsymbol{\sigma}^1$, $\boldsymbol{\sigma}^2$, $\boldsymbol{\sigma}^3$ and $\boldsymbol{\Sigma}^1$, $\boldsymbol{\Sigma}^2$, $\boldsymbol{\Sigma}^3$, are also all present, making the equation significantly different in its *physical* meaning to the one that we had before the transformation. This is now truly the story of two spaces, a fact which was completely lost in the original equation.

We could continue developing this form of the equation, but the physical meaning will become even more apparent if we replace our 5 Pauli generators by their equivalent from the double vector algebra or, because of its origin in the combination of space, time, mass and charge, the vector-quaternion algebra. This will become even more convenient when we realise that the γ matrices, or their algebraic equivalents, are to be found within the wavefunction ψ, as well as in the operator. Of course, the signs of the quaternion units will be arbitrary, so, purely for convenience, we can replace the γ or double Pauli representations by, say,

$$-i\boldsymbol{k},\ -\boldsymbol{i}\mathbf{i},\ -\boldsymbol{i}\mathbf{j},\ -\boldsymbol{i}\mathbf{k},\ i\boldsymbol{j}$$

A convenient way would be to simultaneously replace $\boldsymbol{\sigma}^1$, $\boldsymbol{\sigma}^2$, $\boldsymbol{\sigma}^3$ in the second $\boldsymbol{\sigma} - \boldsymbol{\Sigma}$ equation by $\mathbf{i}$, $\mathbf{j}$, $\mathbf{k}$ and $\boldsymbol{\Sigma}^1$, $\boldsymbol{\Sigma}^2$, $\boldsymbol{\Sigma}^3$ by $\boldsymbol{i}$, $\boldsymbol{j}$, $\boldsymbol{k}$. The unit matrices cancel as the whole expression is equated to 0.

Our equation now becomes

$$\left(-\boldsymbol{k}\frac{\partial}{\partial t} - i\boldsymbol{i}\mathbf{i}\frac{\partial}{\partial x} - i\boldsymbol{i}\mathbf{j}\frac{\partial}{\partial y} - i\boldsymbol{i}\mathbf{k}\frac{\partial}{\partial y} + \boldsymbol{j}m\right)\psi = \left(-\boldsymbol{k}\frac{\partial}{\partial t} - i\boldsymbol{i}\nabla + \boldsymbol{j}m\right)\psi = 0$$

and we can see that it represents a *purely mathematical* development from the original version, without any additional physical assumption. However, we obtain a major physical result as soon as we insert a plane wave solution, for a free particle, say,

$$\psi = Ae^{-i(Et-\mathbf{p}\cdot\mathbf{r})},$$

into this equation, for we immediately obtain

$$\left(-\boldsymbol{k}\frac{\partial}{\partial t} - i\boldsymbol{i}\nabla + \boldsymbol{j}m\right)Ae^{-i(Et-\mathbf{p}\cdot\mathbf{r})} = (i\boldsymbol{k}E + \boldsymbol{i}\mathbf{p} + \boldsymbol{j}m)Ae^{-i(Et-\mathbf{p}\cdot\mathbf{r})} = 0.$$

which has exactly the same form as the equation we obtained directly from the parameter group in the previous chapter, and which we can recognise as expressing a minimal statement of the variation of space and time within a structure created from the minimal algebraic generators of the parameter group.

We can see also that, if energy is conserved, the term in the bracket on the right-hand side is a nilpotent or square root of 0, and that the equation can only hold if A becomes identical to $(i\boldsymbol{k}E + \boldsymbol{i}\mathbf{p} + \boldsymbol{j}m)$ or a scalar multiple of it, which means that ψ is also a nilpotent. (Here, of course, we need to emphasize that the multivariate properties of $\mathbf{p}$ allow us to use the 'spin' terms $\mathbf{p}$ and ∇ instead of the 'helicity' terms $\boldsymbol{\sigma}\cdot\mathbf{p}$ and $\boldsymbol{\sigma}\cdot\boldsymbol{\nabla}$, where $\boldsymbol{\sigma}$ is a unit pseudovector of magnitude $-\mathbf{1}$, in a nilpotent structure, since $(\boldsymbol{\sigma}\cdot\mathbf{p})^2 = \mathbf{pp} = p^2$.) The very remarkable thing about this representation, and it carries over into cases where the fermion is not a free particle, is that the wavefunction is an explicit expression involving energy and momentum terms, $(i\boldsymbol{k}E + \boldsymbol{i}\mathbf{p} + \boldsymbol{j}m)e^{-i(Et-\mathbf{p}\cdot\mathbf{r})}$, a physical *object*, not a black box hidden behind the symbol ψ, which can only be seen as a calculating device. Quantum mechanics remains abstract and probabilistic, but it ceases to look arbitrary. We need to think outside the black box!

Of course, the spinor properties of the algebra still hold, even when we don't use a matrix representation, and ψ will not actually be a single term but rather a 4-component spinor, incorporating fermion/antifermion and spin up/down states. We can easily identify these possibilities with the arbitrary sign options for the iE and $\mathbf{p}$ (or $\boldsymbol{\sigma}\cdot\mathbf{p}$) terms, and this is immediately accommodated in the nilpotent formalism by transforming $(i\boldsymbol{k}E + \boldsymbol{i}\mathbf{p} + \boldsymbol{j}m)$ into a column vector with four sign combinations of iE and $\mathbf{p}$, which may be written in abbreviated form as $(\pm i\boldsymbol{k}E \pm \boldsymbol{i}\mathbf{p} + \boldsymbol{j}m)$. Written out in full the four components are:

$$\begin{gathered}(i\boldsymbol{k}E + \boldsymbol{i}\mathbf{p} + \boldsymbol{j}m)\\(i\boldsymbol{k}E - \boldsymbol{i}\mathbf{p} + \boldsymbol{j}m)\\(-i\boldsymbol{k}E + \boldsymbol{i}\mathbf{p} + \boldsymbol{j}m)\\(-i\boldsymbol{k}E - \boldsymbol{i}\mathbf{p} + \boldsymbol{j}m)\end{gathered}$$

The signs are, of course, intrinsically arbitrary, but it is convenient to identify the four states by adopting a convention, say,

$(i\boldsymbol{k}E + \boldsymbol{i}\mathbf{p} + \boldsymbol{j}m)$	fermion spin up
$(i\boldsymbol{k}E - \boldsymbol{i}\mathbf{p} + \boldsymbol{j}m)$	fermion spin down
$(-i\boldsymbol{k}E + \boldsymbol{i}\mathbf{p} + \boldsymbol{j}m)$	antifermion spin down
$(-i\boldsymbol{k}E - \boldsymbol{i}\mathbf{p} + \boldsymbol{j}m)$	antifermion spin up

Once the convention is decided, however, the spin state of the particle (or, more conventionally, the helicity or handedness $\boldsymbol{\sigma}\cdot\mathbf{p}$, the direction of spin relative to motion, with right-handed being positive) is fixed by the ratio of the signs of E and $\mathbf{p}$. So $\boldsymbol{i}\mathbf{p}/i\boldsymbol{k}E$ has the same helicity as $(-\boldsymbol{i}\mathbf{p})/(-i\boldsymbol{k}E)$, but the opposite helicity to $\boldsymbol{i}\mathbf{p}/(-i\boldsymbol{k}E)$. In the nilpotent theory, the first or lead term represents the particle as observed, though all real fermions incorporate both spin states. If we wanted to represent an antifermion spin down, that term would become the lead term, and all the remaining terms would follow the same cycle of sign changes.

Conventionally, relativistic quantum mechanics uses the large matrix equation (1), but this has massive disadvantages, apart from the difficulty in writing down the amplitudes and the hidden nature of the terms in the wavefunction. First of all, the γ matrices are not all real or all complex, so they don't have the same mathematical status and don't look symmetric — this is because we are trying to use a 2-dimensional algebra (complex numbers) to represent a 3-dimensional system. For the same reason, they are not symmetric as a layout, γ^3 being skewed with respect to the other two. These have damaging consequences, for example, creating the expression $\partial/\partial x + i\partial/\partial$y, which denies the rotational symmetry of the momentum operator and simply has no counterpart in nature. A similar problem happens with the energy operator $\partial/\partial t$, which is variously + and – with respect to im. This carries through at higher levels and means that we have to duplicate equations with alternative solutions, sometimes with a broken connection between them. In addition, we have an enormous number of terms, 16 in the operator and 4 in the wavefunction, making 64 in total, which we have to consider at once. All this comes about because we haven't respected nature's choice of mathematics. If we do, then the result looks very different.

Using the quaternion operators, $\boldsymbol{i}$, $\boldsymbol{j}$, $\boldsymbol{k}$, almost as an extension of the way we use complex numbers to separate out the different parts into their appropriate compartments, we can unscramble or defragment the equation, so that, instead of looking like the left-hand side of our diagram, it looks more like the right, with energy, momentum and mass terms in both operator and amplitude (represented here by the differently coloured bands), collected under the respective labels $\boldsymbol{k}$, $\boldsymbol{i}$ and $\boldsymbol{j}$:

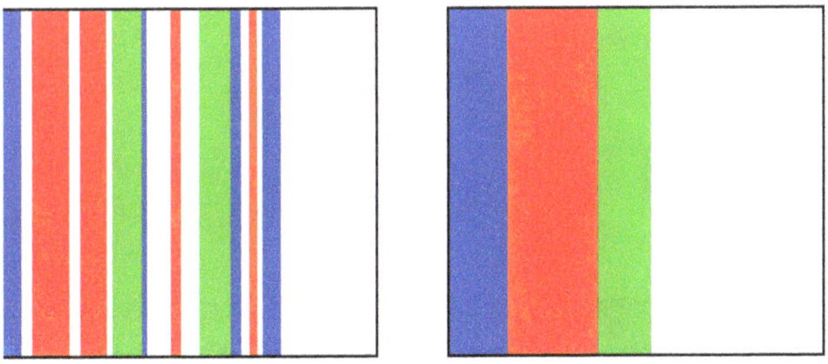

$$\left(-\boldsymbol{k}\frac{\partial}{\partial t} - i\boldsymbol{i}\nabla + \boldsymbol{j}m\right)(\pm i\boldsymbol{k}E \pm \boldsymbol{i}\mathbf{p} + \boldsymbol{j}m)e^{i(Et \pm \mathbf{p}\cdot\mathbf{r})} = 0$$

Here, we have given the full specification for the spinor wavefunction, with four amplitudes and four corresponding phase factors. But, we can also make another massive simplification. This is best seen by writing out the equation for each of the terms separately.

$$\left(-\boldsymbol{k}\frac{\partial}{\partial t} - \boldsymbol{i}\boldsymbol{i}\nabla + \boldsymbol{j}m\right)(\boldsymbol{ik}E + \boldsymbol{i}\mathbf{p} + \boldsymbol{j}m)e^{-i(Et-\mathbf{p}\cdot\mathbf{r})} = 0$$

$$\left(-\boldsymbol{k}\frac{\partial}{\partial t} - \boldsymbol{i}\boldsymbol{i}\nabla + \boldsymbol{j}m\right)(\boldsymbol{ik}E - \boldsymbol{i}\mathbf{p} + \boldsymbol{j}m)e^{-i(Et+\mathbf{p}\cdot\mathbf{r})} = 0$$

$$\left(-\boldsymbol{k}\frac{\partial}{\partial t} - \boldsymbol{i}\boldsymbol{i}\nabla + \boldsymbol{j}m\right)(-\boldsymbol{ik}E + \boldsymbol{i}\mathbf{p} + \boldsymbol{j}m)e^{i(Et+\mathbf{p}\cdot\mathbf{r})} = 0$$

$$\left(-\boldsymbol{k}\frac{\partial}{\partial t} - \boldsymbol{i}\boldsymbol{i}\nabla + \boldsymbol{j}m\right)(-\boldsymbol{ik}E - \boldsymbol{i}\mathbf{p} + \boldsymbol{j}m)e^{i(Et-\mathbf{p}\cdot\mathbf{r})} = 0$$

Here, we have one operator, four amplitude terms and four phase factors. But the reduction of the operator down from sixteen terms to one has given us the 'logical' space to transfer the variation in the *phase factor* to the operator, so extending the operator to four terms (written out in a row vector), operating on four amplitudes (written out in a column vector) with a single shared phase. Written out as separate equations, this would become:

$$\left(-\boldsymbol{k}\frac{\partial}{\partial t} - \boldsymbol{i}\boldsymbol{i}\nabla + \boldsymbol{j}m\right)(\boldsymbol{ik}E + \boldsymbol{i}\mathbf{p} + \boldsymbol{j}m)e^{-i(Et-\mathbf{p}\cdot\mathbf{r})} = 0$$

$$\left(-\boldsymbol{k}\frac{\partial}{\partial t} + \boldsymbol{i}\boldsymbol{i}\nabla + \boldsymbol{j}m\right)(\boldsymbol{ik}E - \boldsymbol{i}\mathbf{p} + \boldsymbol{j}m)e^{-i(Et-\mathbf{p}\cdot\mathbf{r})} = 0$$

$$\left(\boldsymbol{k}\frac{\partial}{\partial t} - \boldsymbol{i}\boldsymbol{i}\nabla + \boldsymbol{j}m\right)(-\boldsymbol{ik}E + \boldsymbol{i}\mathbf{p} + \boldsymbol{j}m)e^{-i(Et-\mathbf{p}\cdot\mathbf{r})} = 0$$

$$\left(\boldsymbol{k}\frac{\partial}{\partial t} + \boldsymbol{i}\boldsymbol{i}\nabla + \boldsymbol{j}m\right)(-\boldsymbol{ik}E - \boldsymbol{i}\mathbf{p} + \boldsymbol{j}m)e^{-i(Et-\mathbf{p}\cdot\mathbf{r})} = 0$$

Since finding the phase factor is the biggest single problem in quantum mechanics, then we can expect a very significant increase in calculating power from this modification alone. Using again a compactified notation for row and column, this set of equations can be written:

$$\underset{row}{(\mp\boldsymbol{k}\tfrac{\partial}{\partial t} \mp \boldsymbol{i}\boldsymbol{i}\nabla + \boldsymbol{j}m)}\ \underset{column}{(\pm\boldsymbol{ik}E \pm \boldsymbol{i}\mathbf{p} + \boldsymbol{j}m)}e^{-i(Et-\mathbf{p}\cdot\mathbf{r})} = 0$$

In this form of the equation, operator and amplitude look the same, especially if we write the first bracket in 'operator' form, with $i\partial/\partial t \to E$

and $-i\nabla \to \mathbf{p}$:

$$(\pm ikE \pm i\mathbf{p} + jm)(\pm ikE \pm i\mathbf{p} + jm)e^{-i(Et-\mathbf{p}\cdot\mathbf{r})} = 0$$

Both are 4-component spinors. We note also that the Feynman prescription (obtainable from the Klein-4 group symmetry) of antifermions having reversed time is exactly obeyed. It seems to me that this is the form of the equation we would have arrived at if we could have derived it from first principles from the parameter group, using all available sign variations, rather than by 'reverse engineering' from the combination of the separately derived theories of special relativity and nonrelativistic quantum mechanics. It is apparent also that the nilpotent expression

$$(\pm ikE \pm i\mathbf{i}p_x \pm i\mathbf{j}p_y \pm i\mathbf{k}p_z + jm)(\pm ikE \pm i\mathbf{i}p_x \pm i\mathbf{j}p_y \pm i\mathbf{k}p_z + jm) = 0$$

or

$$(\pm ikE \pm i\mathbf{p} + jm)(\pm ikE \pm i\mathbf{p} + jm) = 0$$

can be used to create the Klein–Gordon equation that applies to bosons as well as fermions, by treating the E and $\mathbf{p}$ terms in *both* brackets as operators, acting on a particle with wavefunction $A\, e^{-i(Et-\mathbf{p}\cdot\mathbf{r})}$, leading to

$$\left(\mp k\frac{\partial}{\partial t} \mp ii\nabla + jm\right)\left(\mp k\frac{\partial}{\partial t} \mp ii\nabla + jm\right) Ae^{-i(Et-\mathbf{p}\cdot\mathbf{r})} = 0$$

or

$$\left(\nabla - \frac{\partial}{\partial t} - m\right) Ae^{-i(Et-\mathbf{p}\cdot\mathbf{r})} = 0.$$

It is clear that the Klein–Gordon equation will apply to any particle with phase factor $e^{-i(Et-\mathbf{p}\cdot\mathbf{r})}$, as it leads to the product $(\pm ikE \pm i\mathbf{p} + jm)$ $(\pm ikE \pm i\mathbf{p} + jm)$. However, the Dirac equation will not apply to a particle with a scalar amplitude, as application of the single differential operator will only produce one factor containing $(\pm ikE \pm i\mathbf{p} + jm)$, and so be unable to zero the result.

5.3 Using discrete differentiation

With a further refinement, we can reduce the amount of information needed to specify relativistic quantum mechanics to just a two-term operator while reducing the 'wavefunction' to an amplitude without a phase factor. Here, we use a discrete or anticommutative differentiation process with a correspondingly discrete wavefunction. We start by defining a discrete differentiation (mainly following Kauffman[9]) of the function F, which preserves the Leibniz chain rule, by taking:

$$\frac{\partial F}{\partial t} = [F, \mathcal{H}] = [F, E] \quad \text{and} \quad \frac{\partial F}{\partial X_i} = [F, P_i],$$

with $\mathcal{H} = E$ and P_i representing energy and momentum operators respectively. Taking the case where velocity operators are not in evidence, we use $\partial F/\partial t$ rather than Kauffman's dF/dt. The mass term (which, of course, has only a passive role in quantum mechanics) disappears in the operator, though it is retained in the amplitude.

Suppose we define a nilpotent amplitude

$$\psi = \boldsymbol{ik}E + \boldsymbol{i}\mathbf{i}P_1 + \boldsymbol{i}\mathbf{j}P_2 + \boldsymbol{i}\mathbf{k}P_3 + \boldsymbol{j}m$$

and an operator

$$\mathcal{D} = \boldsymbol{ik}\frac{\partial}{\partial t} + \boldsymbol{i}\mathbf{i}\frac{\partial}{\partial X_1} + \boldsymbol{i}\mathbf{j}\frac{\partial}{\partial X_2} + \boldsymbol{i}\mathbf{k}\frac{\partial}{\partial X_3},$$

with

$$\frac{\partial \psi}{\partial t} = [\psi, \mathcal{H}] = [\psi, E] \quad \text{and} \quad \frac{\partial \psi}{\partial X_i} = [\psi, P_i],$$

After some basic algebraic manipulation, we obtain

$$\begin{aligned}-\mathcal{D}\psi &= i\psi(\boldsymbol{ik}E + \boldsymbol{i}\mathbf{i}P_1 + \boldsymbol{i}\mathbf{j}P_2 + \boldsymbol{i}\mathbf{k}P_3 + \boldsymbol{j}m) \\ &\quad + i(\boldsymbol{ik}E + \boldsymbol{i}\mathbf{i}P_1 + \boldsymbol{i}\mathbf{j}P_2 + \boldsymbol{i}\mathbf{k}P_3 + \boldsymbol{j}m)\psi \\ &\quad - 2i(E^2 - P_1^2 - P_2^2 - P_3^2 - m^2).\end{aligned}$$

When ψ is nilpotent, then

$$\mathcal{D}\psi = \left(\boldsymbol{ik}\frac{\partial}{\partial t} + \boldsymbol{i}\nabla\right)\psi = 0.$$

Generalising to four states, with $\mathcal{D}$ and ψ represented as 4-spinors, then

$$\mathcal{D}\psi = \left(\pm\boldsymbol{ik}\frac{\partial}{\partial t} \pm \boldsymbol{i}\nabla\right)(\pm\boldsymbol{ik}E \pm \boldsymbol{i}\mathbf{i}P_1 \pm \boldsymbol{i}\mathbf{j}P_2 \pm \boldsymbol{i}\mathbf{k}P_3 + \boldsymbol{j}m) = 0$$

becomes an equivalent of the nilpotent Dirac equation in this discrete calculus. The reason why this derivation is important is that it does not require the $\pm i$ (or $i\hbar$) term usually applied to the differential operators in canonical quantization, though, because our expression incorporates all four sign variations, we can include this if we want to without changing anything significant in the equation. This means that we can make a smooth transition between classical and quantum conditions, which points to an important connection between mass and the appearance of $\pm i$ terms in quantum mechanical operators and amplitudes.

5.4 Idempotents and nilpotents

Relativistic quantum mechanics was always assumed to require idempotent, rather than nilpotent wavefunctions, essentially because spinors are built up from primitive idempotents. Idempotents square to themselves rather than to zero, but, as we saw in Chapter 2, they are closely related to nilpotents, and each can be converted into the other using quaternion multipliers. We can, in fact, make *exactly the same equation* look either idempotent or nilpotent simply by redistributing a single algebraic unit between the sections of the equation defined as operator and wavefunction. So, we can write the basic equation as either

$$[(\mp \boldsymbol{k}\partial/\partial t \mp \boldsymbol{i}\mathbf{i}\nabla + \boldsymbol{j}m)\boldsymbol{j}][\boldsymbol{j}(\pm i\boldsymbol{k}E \pm \boldsymbol{i}\mathbf{p} + \boldsymbol{j}m)e^{-i(Et-\mathbf{p}\cdot\mathbf{r})}] = 0$$

operator *idempotent wavefunction*

or

$$[(\mp \boldsymbol{k}\partial/\partial t \mp \boldsymbol{i}\mathbf{i}\nabla + \boldsymbol{j}m)\boldsymbol{j}\boldsymbol{j}][(\pm i\boldsymbol{k}E \pm \boldsymbol{i}\mathbf{p} + \boldsymbol{j}m)e^{-i(Et-\mathbf{p}\cdot\mathbf{r})}] = 0$$

operator *nilpotent wavefunction*

without changing either physics or mathematics. The wavefunction, in fact, has both idempotent and nilpotent properties. Even the wavefunction in the original, 'conventional' version of Equation (1) was a nilpotent, because we brought out the nilpotent characteristics merely by multiplying the operator from the left, and not by changing the wavefunction. The idempotent aspect ensures that the 4-component wavefunction acts as a spinor, as in the conventional version of relativistic quantum mechanics.[a]

Besides helping to define spinor characteristics, the idempotent property, as we will show, has an additional special physical significance, related to vacuum. However, the nilpotent property has more powerful physical characteristics. Quantum mechanics is not just a set of mathematical formalisms which act as calculating devices. Each formalism tells us

[a]The four idempotent terms in the column vector $\boldsymbol{j}(\pm i\boldsymbol{k}E \pm \boldsymbol{i}\mathbf{p} + \boldsymbol{j}m)$ sum up to unity after normalization and, if combined with the appropriate quaternionic coefficients identifying their positions in the column, produce zero products between any two terms as required (which, in principle, demonstrates their intrinsic nilpotency). The nilpotent vector $(\pm i\boldsymbol{k}E \pm \boldsymbol{i}\mathbf{p} + \boldsymbol{j}m)$ also has this intrinsic idempotent property, by absorbing the initial factor $\boldsymbol{j}$ into the quaternion coefficients.

something about the nature of the physical system, and, as we will show, nilpotency is a statement of a *physical* principle, or one that is perhaps even more universal, rather than just a mathematical operation. As we will soon show, the nilpotent formalism is optimised for calculating efficiency, even by comparison with the nonrelativistic versions of quantum mechanics. One of the main reasons for this is that the reduction to a single phase factor means that all aspects of a problem can be dealt with in a single calculation, but a more important one is that the extra constraint of nilpotency immediately forces on us an interpretation which connects every fermion holistically and nonlocally with the rest of the universe, in a way that makes much of the formal apparatus of relativistic quantum mechanics redundant. It seems to be impossible to make quantum physics any more minimal.

5.5 Pauli exclusion

If nilpotency is physically significant, what does it actually mean? We can answer this immediately: Pauli exclusion. A fermion, with a nilpotent wavefunction, let's say ψ_1, is automatically Pauli exclusive, because the combination state it forms with another fermion in an identical state, $\psi_1\psi_1$, will necessarily be zero. Naturally occurring fermions, however, are not free particles, and the breakthrough occurs when we say that Pauli exclusion will occur for non-free fermions *for exactly the same reasons.* In these cases, we can redefine the operators E and $\mathbf{p}$ to incorporate any number of field terms or covariant derivatives, so that E now becomes, say, $i\partial/\partial t + e\phi + \cdots$, and $\mathbf{p}$ becomes, say, $-i\nabla + e\mathbf{A} + \cdots$, while preserving the overall structure of the nilpotent operator as $(\pm i\boldsymbol{k}E \pm \boldsymbol{i}\mathbf{p} + \boldsymbol{j}m)$.

Clearly, if we do this, the E and $\mathbf{p}$ interpreted as eigenvalues in the amplitude will require more complicated expressions due to the presence of the additional terms in the operator. The phase factor will also no longer be the $e^{-i(Et-\mathbf{p}\cdot\mathbf{r})}$ that we generate for the free particle; but the system will still be constrained by the need to maintain Pauli exclusion for all fermions, whether free or interacting, and its properties will be uniquely determined by the information that the operator has encoded. This will be true even if the external field terms are defined by expectation values, as with the Lamb shift, or in terms of quantum fields.

Writing an operator in the generic form $(\pm i\boldsymbol{k}E \pm \boldsymbol{i}\mathbf{p} + \boldsymbol{j}m)$, with E and $\mathbf{p}$ now defined as generic terms involving differentials and associated potentials, means that we immediately specify the entire quantum mechanics of the system. The wavefunction and even *the equation itself* become redundant as independent sources of information. The operator alone uniquely and completely determines the phase factor which will create a nilpotent amplitude. This does not require a special quantum mechanical equation, only a secondary functional one:

$$(\text{operator acting on phase factor})^2 = 0.$$

Even the spinor representation loses its independent status, as the first of the four terms, say $(i\boldsymbol{k}E+\boldsymbol{i}\mathbf{p}+\boldsymbol{j}m)$, will uniquely and completely specify the remaining three by automatic sign variation. (For this reason, we will often specify the operator in an abbreviated form using only the first term, and, when we write down an amplitude, we will often leave out the phase factor, which again contains no independent information.) Instead of needing an operator with 16 terms and an amplitude with 4 terms, we can specify the entire system with just a single term from an operator. Astonishingly, we don't need an equation at all to do quantum mechanics. The seemingly most fundamental structure in physics as we know it today is not an equation but an expression linking the symmetric algebra of four basic quantities, and the abstract properties that these algebras create, which we interpret as 'physics'.

5.6 Vacuum

Vacuum is an important concept in quantum physics but it is not well understood. In the nilpotent theory, however, it has a very clear meaning, which allows an immediate transition to quantum field theory, without any formal process of second quantization. It also invokes the fundamental idea of totality zero. We imagine creating a fermion in some particular state (determined by added potentials, interaction terms, *etc.*) *ab initio*, that is, from absolutely nothing, a complete void or totality zero. We can then imagine vacuum as what is left in *nothing* — that is, everything other than the fermion. If we define the wavefunction of the fermion as, say, ψ_f, the wavefunction of vacuum will become $\psi_v = -\psi_f$. The superposition of fermion and vacuum will be the initial zero state, $\psi_f + \psi_v = \psi_f - \psi_f = 0$, and, because the fermion is a nilpotent, the combination state

$$\psi_f\psi_v = -\psi_f\psi_f = -(\pm i\boldsymbol{k}E \pm \boldsymbol{i}\mathbf{p} + \boldsymbol{j}m)(\pm i\boldsymbol{k}E \pm \boldsymbol{i}\mathbf{p} + \boldsymbol{j}m)$$

will also be zero. So, vacuum can be described as the 'hole' in the totality zero state produced by the creation of the fermion, and this becomes the 'rest of the universe' that the fermion sees and interacts with. So, if we want to 'create' a fermion with interacting field terms, then the 'rest of the universe' has to be 'constructed' simultaneously to make the existence of a fermion in that state possible.

Vacuum in this definition leads to a view of the universe as a zero totality, which, at the creation of every new fermionic state, divides into two parts — the local fermionic state and the nonlocal vacuum. These are connected with the simultaneous existence of two vector spaces to create the mathematical structure: the real space with units **i**, **j**, **k** and the vacuum space with units **i**, **j**, **k** (derived from the mass–time–charge units of 1, i, $\boldsymbol{i}$, $\boldsymbol{j}$, $\boldsymbol{k}$). The nilpotent formalism indicates that a fermion 'constructs' its own vacuum, or the entire 'universe' in which it operates, and we can consider the vacuum to be 'delocalised' to the extent that the fermion is 'localised'. We can consider the nilpotency as defining the interaction between the localised fermionic state and the delocalised vacuum, with which it is uniquely self-dual, the phase being the mechanism through which this is accomplished. We can also consider Pauli exclusion as saying that no two fermions can share the same vacuum.

Nilpotency now makes it possible for us to understand nonlocality, giving us a simple separation of local and nonlocal. The 'local' (which implies Lorentz or relativistic invariance) is defined as whatever happens inside the nilpotent structure $(\pm i\boldsymbol{k}E \pm \boldsymbol{i}\mathbf{p} + \boldsymbol{j}m)$, and the 'nonlocal' as whatever happens outside it. The bracketed term representing the fermion creation operator or wavefunction determines how conservation of energy applies to that fermion, as squaring the wavefunction and equating to zero gives us back the energy–momentum equation, and, of course, it is local, as the required Lorentzian structure is intrinsic. However, the addition and multiplication of nilpotent wavefunctions construct the nonlocal processes of superposition and combination, and these processes do not require a Lorentzian structure.

No single fermion can ever be isolated but must be interacting, and construct a 'space', so that its vacuum is not localised on itself. A point-like fermion necessarily requires a dispersed vacuum. Vacuum is intrinsically nonlocal. Because the fermion is localised, then the rest of the universe is necessarily nonlocalised. If the fermion is a point, as experiments suggest that it may be, then the rest of the universe is defined as everything outside that point. We can't define a single (noninteracting) fermion to exist — it can only be defined if we also define its vacuum. So the nonlocal connection which makes Pauli exclusion possible can be thought to occur through the vacua for each fermion.

Nilpotency isn't the usual way of expressing Pauli exclusion mathematically. In the standard interpretation, wavefunctions or amplitudes are also

Pauli exclusive because they are antisymmetric, with nonzero

$$\psi_1\psi_2 - \psi_2\psi_1 = -(\psi_2\psi_1 - \psi_1\psi_2)$$

This, however, is automatic in the nilpotent formalism, where the expression becomes

$$\begin{aligned}&(\pm i\boldsymbol{k}E_1 \pm \boldsymbol{i}\mathbf{p}_1 + \boldsymbol{j}m_1)(\pm i\boldsymbol{k}E_2 \pm \boldsymbol{i}\mathbf{p}_2 + \boldsymbol{j}m_2)\\ &\quad -(\pm i\boldsymbol{k}E_2 \pm \boldsymbol{i}\mathbf{p}_2 + \boldsymbol{j}m_2)(\pm i\boldsymbol{k}E_1 \pm \boldsymbol{i}\mathbf{p}_1 + \boldsymbol{j}m_1)\\ &= 4\mathbf{p}_1\mathbf{p}_2 - 4\mathbf{p}_2\mathbf{p}_1 = 8i\mathbf{p}_1 \times \mathbf{p}_2.\end{aligned}$$

which is clearly antisymmetric. But the result also tells us something new, for it requires a nilpotent wavefunction to have a $\mathbf{p}$ vector in spin space at a different orientation to any other. The instantaneous nonlocal correlation of all nilpotent wavefunctions could then require the intersection of the planes corresponding to all the different $\mathbf{p}$ vector directions. We can even consider these intersections as actually creating the *meaning* of Euclidean space, with an intrinsic spherical symmetry generated by the fermions themselves.

It seems likely that the universal collection of spin axis directions at any given instant is not repeatable, giving a unique direction for time, and — including energy as the 'time' component — making fermionic world-lines unique as well. A further connection with irreversibility or a unique time direction is suggested by the fact that an interaction between fermions with differently oriented vectors $\mathbf{p}_1$ and $\mathbf{p}_2$ will produced a reduced magnitude of momentum within the system unless the even more 'organized' amount of rest mass is reduced. For nilpotents, $\mathbf{p}_1$ and $\mathbf{p}_2$ can never be parallel, and the vector sum of components in different directions will always be less than the scalar magnitudes added. In either case, the system will become less coherent or more entropic, as required by the second law of thermodynamics, the only law of physics which requires an irreversibility of time.

Pauli exclusion means that defining a fermion implies simultaneous definition of vacuum as 'the rest of the universe' with which it interacts. The nilpotent structure then provides energy–momentum conservation without requiring the system to be closed, since the E and $\mathbf{p}$ terms also contain all possible interactions. Again, we see that the nilpotent structure is naturally thermodynamic, and provides a route to a mathematization of nonequilibrium thermodynamics — all systems in this formulation are open systems. Also, the formation of any new state, which is determined by the nature of all other nilpotent states, is a creation event within a unique birth-ordering. Each 'creation' event (which includes any interaction and any change in parameters, as well as entirely new fermionic creations) also necessarily changes all existing states to some degree. In this sense, a nilpotent structure uniquely allows us to conceive of the infinite while only observing the finite.

We now have at least five different meanings for the expression

$$(\pm \boldsymbol{ik}E \pm \mathbf{ip} + \boldsymbol{j}m)(\pm \boldsymbol{ik}E \pm \mathbf{ip} + \boldsymbol{j}m)\phi \to 0$$

with ϕ an (optional) arbitrary scalar factor (phase, *etc.*):

Expression	Meaning
classical	special relativity
operator × operator	Klein–Gordon equation
operator × wavefunction	Dirac equation
wavefunction × wavefunction	Pauli exclusion
fermion × vacuum	thermodynamics

It is characteristic of a theory in which duality is so deeply embedded to create such multiple meanings. The same is true of $\boldsymbol{i}$, $\boldsymbol{j}$, $\boldsymbol{k}$, which, as we will see, have many meanings, including charge operators, vacuum operators and generators of P, C and T symmetry transformations.

Yet another way of looking at Pauli exclusion is to say that the phase factors of all fermion wavefunctions must be unique. In effect, this means that the E and p values, and amplitudes are all unique, which is the same as saying that they are nilpotent. The blue colour indicates that p is in the real space, as oppose to the red vectors which are in the vacuum space. Here, we might return to our dual spaces, i, j, k and i, j, k, which can be said, in one interpretation, to carry the entire information relating to physics, the first being rotationally symmetric and the second rotationally asymmetric. The first is the real space or the space of measurement, and the second the vacuum space or space of interaction. The nilpotent condition requires these to be dual in terms of all physical information, although presenting it in quite different forms except in the construction of spinors. The nilpotency seen in vacuum space, which creates the unique phase factors for fermions, might present this phase as a unique direction on a set of axes defined by the values of E, p and m, by analogy with the unique direction of p on the axes of real space that we derived from the antisymmetric wavefunctions. The mapping of the unique phase onto both spin vector (using i, j, k) and to the axes whose directions were derived from i, j, k, gives a good indication that the nilpotent structure is rightly considered an angular momentum operator, and the link of i, j, k with the three charge components shows the connection between angular momentum and charge which we discussed in Chapter 4.

A number of significant results emerge automatically from the i, j, k or $\mathbf{i}$, $\mathbf{j}$, $\mathbf{k}$ representation. For example, half of the possibilities on one axis (those with $-m$) would be eliminated automatically (as being in the same direction as those with m), providing clear evidence that invariant mass cannot have two signs. Also eliminated would be fermions with zero m, since the directions would all be along the line $E = p$. In addition, such hypothetical massless particles would be impossible for fermions and antifermions with the same helicity, as E, p has the same direction as $-E$, $-p$.

5.7 Quantum mechanics and the quantum field

The nilpotent operator can be used to do ordinary relativistic quantum mechanics, once we have defined a probability density for a nilpotent wavefunction, $(\pm \boldsymbol{ik}E \pm \boldsymbol{i}\mathbf{p} + \boldsymbol{j}m)$. In standard quantum mechanics, we create this by multiplying the *wavefunction* by its complex conjugate. Here, we use the *complex quaternion conjugate* $(\pm \boldsymbol{ik}E \mp \boldsymbol{i}\mathbf{p} - \boldsymbol{j}m)$ (the extra 'quaternion' resulting from the fact that the nilpotent wavefunction differs from a conventional one through premultiplication by a quaternion operator). The unit probability density is then defined by

$$\frac{(\pm \boldsymbol{ik}E \pm \boldsymbol{i}\mathbf{p} + \boldsymbol{j}m)}{\sqrt{2E}} \frac{(\pm \boldsymbol{ik}E \mp \boldsymbol{i}\mathbf{p} - \boldsymbol{j}m)}{\sqrt{2E}} = 1,$$

the $1/\sqrt{2E}$ being a normalizing factor. If we assume that such factors automatically apply in calculations, we can also define $(\pm \boldsymbol{ik}E \mp \boldsymbol{i}\mathbf{p} - \boldsymbol{j}m) = -(\mp \boldsymbol{ik}E \pm \boldsymbol{i}\mathbf{p} + \boldsymbol{j}m)$ as the 'reciprocal' of $(\pm \boldsymbol{ik}E \pm \boldsymbol{i}\mathbf{p} + \boldsymbol{j}m)$.

Rather more significantly, the nilpotent formalism not only creates quantum mechanics, but also implies a full quantum field theory in which the operators act on the entire quantum field, without requiring any formal process of second quantization. The transition to quantum field theory seemingly occurs at the point at which we choose to privilege the operator rather than the equation, and then apply Pauli exclusion to all fermionic states, whether free or bound, regardless of the number of interactions to which they are subject. A nilpotent operator, defined in this way from *absolutely nothing*, then becomes a creation operator acting on vacuum to create the fermion, together with all the interactions in which it is involved. As we have seen, no further mathematical formalism is necessary, and neither quantum mechanics nor quantum field theory requires specification by an equation. Once the operator is defined, the phase factor becomes an expression of all the possible variations in space and time which are encoded in the operator, and is uniquely defined with it. A fermion is thus specified as a set of space and time variations, with the mass term as a purely passive and convenient quantity, rather than necessary information. It would be difficult to imagine anything closer to the ideal which the parameter group embodies.

5.8 Spin and helicity

The derivation of fermion spin $\frac{1}{2}$ illustrates the multivariate nature of the $\mathbf{p}$ operator. The nilpotent operator $(\boldsymbol{k}E + \boldsymbol{i}\mathbf{p} + \boldsymbol{j}m)$ incorporates a Hamiltonian specified as $\mathcal{H} = -i\boldsymbol{k}(\boldsymbol{i}\mathbf{p} + \boldsymbol{j}m) = -i\boldsymbol{j}\mathbf{p} + i\boldsymbol{i}m$. If we *mathematically* define a quantity $\boldsymbol{\sigma} = -\mathbf{1}$ (the pseudovector of magnitude –1 already referred to), then

$$\begin{aligned}[\boldsymbol{\sigma}, \mathcal{H}] &= [-\mathbf{1}, -i\boldsymbol{j}(\mathbf{i}p_1 + \mathbf{j}p_2 + \mathbf{k}p_3) + i\boldsymbol{i}m] = [-\mathbf{1}, -i\boldsymbol{j}(\mathbf{i}p_1 + \mathbf{j}p_2 + \mathbf{k}p_3)] \\ &= 2i\boldsymbol{j}(\mathbf{ij}p_2 + \mathbf{ik}p_3 + \mathbf{ji}p_1 + \mathbf{jk}p_3 + \mathbf{ki}p_1 + \mathbf{kj}p_2) \\ &= -2\boldsymbol{j}(\mathbf{k}(p_2 - p_1) + \mathbf{j}(p_1 - p_3) + \mathbf{i}(p_3 - p_2)) \\ &= -2\boldsymbol{j}\mathbf{1} \times \mathbf{p}.\end{aligned}$$

If $\mathbf{L}$ is an orbital angular momentum $\mathbf{r} \times \mathbf{p}$, then

$$\begin{aligned}[\mathbf{L}, \mathcal{H}] &= [\mathbf{r} \times \mathbf{p}, -i\boldsymbol{j}(\mathbf{i}p_1 + \mathbf{j}p_2 + \mathbf{k}p_3) + i\boldsymbol{i}m] \\ &= [\mathbf{r} \times \mathbf{p}, -i\boldsymbol{j}(\mathbf{i}p_1 + \mathbf{j}p_2 + \mathbf{k}p_3)] \\ &= [\mathbf{r}, -i\boldsymbol{j}(\mathbf{i}p_1 + \mathbf{j}p_2 + \mathbf{k}p_3)] \times \mathbf{p}\end{aligned}$$

But

$$[\mathbf{r}, -i\boldsymbol{j}(\mathbf{i}p_1 + \mathbf{j}p_2 + \mathbf{k}p_3)] = i\mathbf{1}.$$

Hence

$$[\mathbf{L}, \mathcal{H}] = \mathbf{1} \times \mathbf{p},$$

and $\mathbf{L} + \boldsymbol{\sigma}/2$ is a constant of the motion, because

$$[\mathbf{L} + \boldsymbol{\sigma}/2, \mathcal{H}] = 0.$$

In this formalism, the spin $\frac{1}{2}$ term characteristic of fermionic states emerges purely from the multivariate properties of the $\mathbf{p}$ operator, through the additional cross product term with its imaginary coefficient or pseudovector, exactly as Hestenes showed must result from this algebra.[3] Physically, we observe it as an intrinsic angular momentum term requiring a fermion to undergo a 4π, rather than 2π, rotation to return to its starting point. In our interpretation, this comes from the fact that a localised point-like fermion can only be created simultaneously with a mirror image nonlocalised vacuum state. The fermion on its own provides only half of the knowledge we require to specify the system, and this is equivalent to specifying only one of the two spaces. The system spends only half its time as fermion in real space, and the other half as an antifermion in vacuum space, as in the 4-component (spinor) wavefunction. The spin of fermion plus vacuum is, of course, single-valued (0).

We can define helicity ($\boldsymbol{\sigma} \cdot \mathbf{p}$) as another constant of the motion because

$$[\boldsymbol{\sigma}\cdot\mathbf{p}, \mathcal{H}] = [-p, -i\boldsymbol{j}(\mathbf{i}p_1 + \mathbf{j}p_2 + \mathbf{k}p_3) + i\boldsymbol{i}m] = 0$$

We have previously specified that, for a multivariate $\mathbf{p}$,

$$\mathbf{pp} = (\boldsymbol{\sigma} \cdot \mathbf{p})(\boldsymbol{\sigma}\cdot\mathbf{p}) = pp = p^2.$$

So we can also use $\boldsymbol{\sigma} \cdot \mathbf{p}$ ($\boldsymbol{\sigma}\mathbf{p}$) for $\mathbf{p}$ (or $\boldsymbol{\sigma} \cdot \boldsymbol{\nabla}$ ($\boldsymbol{\sigma}\boldsymbol{\nabla}$) for $\boldsymbol{\nabla}$) in the nilpotent operator. As in the graphical representation we have just discussed, a hypothetical fermion/antifermion with zero mass would be reduced to two distinguishable states:

$$(i\boldsymbol{k}E + \boldsymbol{i}\boldsymbol{\sigma} \cdot \mathbf{p} + \boldsymbol{j}m) \to (i\boldsymbol{k}E - \boldsymbol{i}p)$$

$$(-i\boldsymbol{k}E + \boldsymbol{i}\boldsymbol{\sigma} \cdot \mathbf{p} + \boldsymbol{j}m) \to (-i\boldsymbol{k}E - \boldsymbol{i}p)$$

each of which has only a single sign of helicity; $(i\boldsymbol{k}E + \boldsymbol{i}p)$ and $(-i\boldsymbol{k}E + \boldsymbol{i}p)$ are prevented from existing by Pauli exclusion, if we choose the same sign conventions for $\mathbf{p}$. The previous use of $\sigma = -\mathbf{1}$ in deriving spin for states with positive energy suggests that the allowed spin direction for these states must be antiparallel, corresponding to left-handed helicity, with the negative energy, antistates having right-handed helicity. Numerically, $|\pm E| = p$, so the allowed states can be expressed as $\pm E(\boldsymbol{k} - i\boldsymbol{i})$. As in standard theory, if we multiply from the left by the projection operator $(1 - i\boldsymbol{j})/2 \equiv (1 - \gamma^5)/2$ the allowed states will remain unchanged while the excluded ones are zeroed.

Because spin has emerged in this formalism from the specifically multivariate aspect of the operator $\mathbf{p}$, it is necessary to distinguish equations where the space variables are multivariate from those where they are not, as, for example, when polar coordinates are used. If fermions are point particles and their influence is spherically symmetric, then it will be convenient to express the influence of one point-source on another by changing the coordinates of the 'receiving' particle from Cartesian to polar, with the point-particle source at the centre of the coordinate system also defined as the centre of physical influence.

In such cases, an intrinsic spin is no longer structured into the formalism and an *explicit* spin (or total angular momentum) term has to be introduced. Dirac, however, has given a prescription for translating his equation into polar form, where the momentum operator acquires an additional (imaginary) spin (or total angular momentum) term,[8] and we can easily adapt this to represent a polar transformation of the multivariate vector operator:

$$\nabla \to \left(\frac{\partial}{\partial r} + \frac{1}{r}\right) \pm i\frac{j + \frac{1}{2}}{r}.$$

and use this to define a non-time varying nilpotent operator in polar coordinates:

$$(i\boldsymbol{k}E - \boldsymbol{i}i\nabla + \boldsymbol{j}m) \to \left(i\boldsymbol{k}E - \boldsymbol{i}i\left(\frac{\partial}{\partial r} + \frac{1}{r} \pm i\frac{j + \frac{1}{2}}{r}\right) + \boldsymbol{j}m\right).$$

If we take the use of polar coordinates as representing spherical symmetry with respect to a point source, then this operator has no nilpotent solutions unless the $i\boldsymbol{k}E$ term also contains an expression proportional to $1/r$ to cancel those produced by the $\boldsymbol{i}i\nabla$ term. We recognise this as a Coulomb or inverse-square interaction component. In other words, simply defining a point source forces us to assume that a Coulomb interaction component is necessary for any nilpotent fermion defined with respect to it. All known forces have such components, together with an associated $U(1)$ symmetry. For the gravitational and electric forces, it is the main or complete description; for the strong force it is the one-gluon exchange; for the weak field it is the hypercharge and the B^0 gauge field. Its effect is connected purely with scale or magnitude and we can associate it with the coupling constant. In fundamental terms it arises because defining a point in any meaningful way in 3-dimensional space requires a dual space which is structured on the basis of point charges. In addition, the only way of *fixing* a point in a nonconserved space with no identifiable units is to fix it in the space of a conserved quantity which is made to coincide, through nilpotency, with this one. By making this Coulomb component a consequence of nilpotency, we can also see it as a consequence of Pauli exclusion.

5.9 *Zitterbewegung* and Berry phase

Originally, Dirac used α and β operators in his equation, rather than γ and γ^0; α was equivalent to $\gamma\gamma^0$ and β to γ^0. So far, the constants $\hbar$ and c have been left out of our equations, but now it will be convenient to include them. Using these different conventions on symbolism, the nilpotent Hamiltonian becomes

$$\mathcal{H} = -\boldsymbol{i}\boldsymbol{j}c\boldsymbol{\sigma}\cdot\mathbf{p} - \boldsymbol{i}\boldsymbol{i}mc^2 = -\boldsymbol{i}\boldsymbol{j}c\mathbf{1}\mathbf{p} - \boldsymbol{i}\boldsymbol{i}mc^2 = \alpha c\mathbf{p} - \boldsymbol{i}\boldsymbol{i}mc^2.$$

With four separate spin states in the system, $\alpha = -\boldsymbol{i}\boldsymbol{j}\mathbf{1}$ can be taken as a dynamical variable, and $\alpha c = -\boldsymbol{i}\boldsymbol{j}\mathbf{1}c$ defined, in terms of the discrete commutator calculus, as a velocity operator, which, for a free particle, becomes:

$$\mathbf{v} = \dot{\mathbf{r}} = \frac{d\mathbf{r}}{dt} = \frac{1}{i\hbar}[\mathbf{r}, \mathcal{H}] = -\boldsymbol{i}\boldsymbol{j}\mathbf{1}c = c\boldsymbol{\alpha}.$$

Because we are now using a velocity operator, dF/dt must be distinguished from $\partial F/\partial t$. The equation of motion for the velocity operator then becomes:

$$\frac{d\alpha}{dt} = \frac{1}{i\hbar}[\boldsymbol{\alpha}, \mathcal{H}] = \frac{2}{i\hbar}(c\mathbf{p} - \mathcal{H}\boldsymbol{\alpha}).$$

Schrödinger's solution of this well-known result[9] first gave the equation of motion for the fermion from the Dirac equation:

$$\mathbf{r}(t) = \mathbf{r}(0) + \frac{c^2\mathbf{p}}{\mathcal{H}} + \frac{\hbar c}{2i\mathcal{H}}[\boldsymbol{\alpha}(0) - c\mathcal{H}^{-1}\mathbf{p}](exp(2i\mathcal{H}t/h) - 1).$$

Though we have nothing to add to the derivation of this result, the nilpotent interpretation is important for our fundamental view of physics. The equation has classical analogues for all its terms, except the third, which Schrödinger interpreted as predicting a violent oscillatory motion or high-frequency vibration (which he called *zitterbewegung*) of the particle at frequency $\approx 2mc^2/\hbar$, and amplitude $\hbar/2mc$, which is related to the Compton wavelength for the particle and directly determined by the particle's rest mass.

The *zitterbewegung*, which can only be nonzero between states with equal momentum and opposite energy, is interpreted as a switching between the four fermionic states, which incorporate matter and antimatter and two directions of spin. It is certainly a vacuum effect, and we can interpret it as a continual redefinition of the localised fermionic state in relation to the nonlocal vacuum, without which it could not be defined at a point, and so an expression of the necessity of dual vector spaces in the description of a discrete particle. It is, in fact, a direct expression of the duality of these spaces. If two ways of presenting the same information can exist, then they will become equally probable and nature will not privilege one over the other, though one may seem more convenient from our perspective.

Dirac saw *zitterbewegung* as implying that a fermion (or any massive particle) actually propagates along the light cone, oscillating between $+c$ and $-c$ at a frequency which determines its measured mass and momentum. According to this argument, a measured value of velocity can only be found by knowing positions at two different times. To find the instantaneous velocity, you have to reduce the time interval to zero, thus fixing the positions with exact precision, and so making the momentum value completely indeterminate according to Heisenberg uncertainty. The ultimate significance of *zitterbewegung* in this context may be that it locates rest mass as the result of defining a singularity.

Zitterbewegung can thus be seen as an intrinsic aspect of defining a fermion as a point-singularity through the nilpotent structure created by dual vector spaces. This brings us back to our topologies of simply- and multiply-connected spaces and the Berry phase (Chapter 2). Some Berry phase phenomena involve a fermion with half-integral spin subjected to a cyclic adiabatic (i.e. effectively non-dissipative) process becoming single valued (i.e. with integer spin) in the presence of either another fermionic state, for example, an electron (Cooper pairing) or nucleus (Jahn–Teller effect), or an 'environment' whose origin is ultimately fermionic. This last could be, for example a vector potential (Aharonov–Bohm effect) or a magnetic flux line (quantum Hall effect). In each of these cases the Berry phase can be interpreted topologically, as in Chapter 2, with the initial fermion travelling in a space that has changed from being simply- to multiply-connected by incorporating the other fermionic state or environment as a 'singularity'.

As previously mentioned, we can regard the unpaired fermion, defined as a pure physical singularity, as existing in its *own* multiply-connected space and thus naturally becoming a spin $\frac{1}{2}$ particle. Experimental evidence to date suggests that fermions are point-like, and in this sense singularities; excluding anything that might be produced by gravity, and bosons as the products of fermion interactions, they are the only known physical singularities. Now, a physical singularity can only be defined with reference to a nonlocalised phase. The phase factor, as we have seen, contains all the information about the singularity but extends everywhere, and overlaps with all other phase factors. It is this which enables two seemingly isolated singularities to interact, and which allows us to describe such interactions in terms of a quantum field. This will become more explicit when we do specific examples in the next chapter. Information from the dual spaces of one system (*i.e.* potentials or even distortions of its space-time structure) creates changes in the dual spaces of the other, via changes in the E and $\mathbf{p}$ terms of its operator, and, through the phase factor, of its amplitude. Even a pure vector potential (as in the Aharonov–Bohm effect, where it is created by a solenoid with no external field) will alter the $\mathbf{p}$ term and so produce these changes. Under cyclic adiabatic conditions, we can consider the E and p magnitudes of the combination to be equalised as in the formation of the bosonic-type states we will consider in the next chapter.

5.10 *CPT* symmetry

If the lead term in the fermionic column vector defines the fermion type, then we can show that the remaining terms are equivalent to the lead term subjected to the respective symmetry transformations, P, T and C, by pre- and post-multiplication by the quaternion units $\boldsymbol{i}$, $\boldsymbol{j}$, $\boldsymbol{k}$ defining the *vacuum space*:

Parity	P	$\boldsymbol{i}(\pm i\boldsymbol{k}E \pm i\mathbf{p} + \boldsymbol{j}m)\boldsymbol{i} = (\pm i\boldsymbol{k}E \mp i\mathbf{p} + \boldsymbol{j}m)$
Time reversal	T	$\boldsymbol{k}(\pm i\boldsymbol{k}\mathrm{E} \pm i\mathbf{p} + \boldsymbol{j}m)\boldsymbol{k} = (\mp i\boldsymbol{k}E \pm i\mathbf{p} + \boldsymbol{j}m)$
Charge conjugation	C	$-\boldsymbol{j}(\pm i\boldsymbol{k}E \pm i\mathbf{p} + \boldsymbol{j}m)\boldsymbol{j} = (\mp i\boldsymbol{k}E \mp i\mathbf{p} + \boldsymbol{j}m)$

The charge conjugation process could equally be represented by

$$\text{Charge conjugation} \quad C \quad i\boldsymbol{j}(\pm i\boldsymbol{k}E \pm i\mathbf{p} + \boldsymbol{j}m)i\boldsymbol{j} = (\mp i\boldsymbol{k}E \mp i\mathbf{p} + \boldsymbol{j}m),$$

showing its ultimate origin in a vector space where $\mathbf{ij} = i\mathbf{k}$. We can easily show by trying it out that the rules

$$CP \equiv T, \quad PT \equiv C, \quad \text{and} \quad CT \equiv P$$

necessarily apply, as also

$$TCP \equiv CPT \equiv \text{identity}$$

as

$$\boldsymbol{k}(-\boldsymbol{j}(\boldsymbol{i}(\pm i\boldsymbol{k}E \pm \boldsymbol{i}\mathbf{p} + \boldsymbol{j}m)\boldsymbol{i})\boldsymbol{j})\boldsymbol{k}$$
$$= -\boldsymbol{kji}(\pm i\boldsymbol{k}E \pm \boldsymbol{i}\mathbf{p} + \boldsymbol{j}m)\boldsymbol{ijk} = (\pm i\boldsymbol{k}E \pm \boldsymbol{i}\mathbf{p} + \boldsymbol{j}m).$$

It is clear that charge conjugation (or exchange of particle and antiparticle) is effectively defined in terms of parity and time reversal, rather than being an independent operation. This comes from the fact that only space and time are active elements with their variation being the coded information that solely determines the phase factor and the entire nature of the fermionic state, while the mass term (which connects with the charge conjugation transformation) is a passive element, which we have shown can even be excluded from the operator without loss of information. The construction of a nilpotent amplitude effectively requires the loss of a sign degree of freedom in one component, E, $\mathbf{p}$ or m, and the passivity of mass makes it the term to which this will apply.

The CPT theorem is based on a combination of relativity and causality. Relativity says that the square of $(\pm i\boldsymbol{k}E \pm \boldsymbol{i}\mathbf{p})$ or its conjugate, $(\pm i\boldsymbol{k}t \pm \boldsymbol{i}\mathbf{r})$, is an invariant, and it is only when $\boldsymbol{j}m$ or the equivalent $\boldsymbol{j}\tau$ is added that we also get causality. Relativity and causality combined require a structure with $\boldsymbol{k}$, $\boldsymbol{i}$ and $\boldsymbol{j}$ on the same footing, and this is also a requirement for CPT symmetry. C, P and T are the symmetries concerning the algebraic signs of three of the four fundamental parameters, and so the CPT theorem is one of the most fundamental in physics. If the fourth parameter, mass, was not unipolar, there would be an $MCPT$ theorem.

Chapter 6

Nilpotent Quantum Mechanics II

6.1 Bosons

Here, we take our quantum mechanics to the next level of complexity and look at interacting fermions. This chapter will follow three stages. First of all, we will consider the *nonlocal* structures that emerge from the nilpotent formalism, in particular the combination states and superpositions which are required to describe bosons and baryons. Then we will see how these nonlocal structures have *local* consequences. In effect, we will see how local changes *inside* nilpotent brackets can produce the same results as nonlocal changes *outside* them. The local changes involve the creation of local potentials added to the energy and momentum operators. Following these local changes, we will then show how the nilpotent operators as now defined lead to just three possible solutions.

Fermion interactions require bosons, and these differ from fermions in having scalar wavefunctions and not being Pauli exclusive. These facts, while fundamental, have never been explained in fundamental terms. The nilpotent 4-spinor, as we have seen, is composed of the lead term, defining the 'real' particle state and three terms, which are, effectively, the P-, T- and C-transformed versions of this state. They are the possible states into which it could transform without changing the magnitude of its energy or momentum. We can also see them as vacuum 'reflections' of the real particle state, and we will later show how they arise from vacuum operations that can be mathematically defined, through a partitioning of the continuous vacuum into a 3-dimensional vacuum space with each reflection being in one 'dimension' of the space. Now, Pauli exclusion prevents a fermion from forming a combination state with itself, but we can imagine it forming a combination state with each of these vacuum 'reflections', and, if the 'reflection' exists or materialises as a 'real' state, then the combined state can form one of the three classes of bosons or boson-like objects: spin 1 boson, spin 0 boson, or fermion–fermion combination.

A combination of fermion and antifermion with the same spins but opposite helicities will produce a state equivalent to a spin 1 boson. Suppose we take the product of a row vector fermion and a column vector antifermion, both written as columns for convenience:

$$\begin{matrix} \textit{row} & & \textit{column} \\ \begin{pmatrix} i\boldsymbol{k}E + \boldsymbol{i}\mathbf{p} + \boldsymbol{j}m \\ i\boldsymbol{k}E - \boldsymbol{i}\mathbf{p} + \boldsymbol{j}m \\ -i\boldsymbol{k}E + \boldsymbol{i}\mathbf{p} + \boldsymbol{j}m \\ -i\boldsymbol{k}E - \boldsymbol{i}\mathbf{p} + \boldsymbol{j}m \end{pmatrix} & . & \begin{pmatrix} -i\boldsymbol{k}E + \boldsymbol{i}\mathbf{p} + \boldsymbol{j}m \\ -i\boldsymbol{k}E - \boldsymbol{i}\mathbf{p} + \boldsymbol{j}m \\ i\boldsymbol{k}E + \boldsymbol{i}\mathbf{p} + \boldsymbol{j}m \\ i\boldsymbol{k}E - \boldsymbol{i}\mathbf{p} + \boldsymbol{j}m \end{pmatrix} \end{matrix}$$

The antifermion structure reverses the signs of E throughout, and spin reversal changes the sign of $\mathbf{p}$, but the phase factor of both fermion and antifermion components will be, according to our original construction of the nilpotent formalism, the same, dependent on the values of E and $\mathbf{p}$ but not on their signs. Sign variations ensure cancellation of all the terms with quaternion coefficients, so the product is a nonzero scalar. The same result will apply if the spin 1 boson is massless (as is the case with such gauge bosons as photons and gluons). Then we have:

$$\begin{pmatrix} i\boldsymbol{k}E + \boldsymbol{i}\mathbf{p} \\ i\boldsymbol{k}E - \boldsymbol{i}\mathbf{p} \\ -i\boldsymbol{k}E + \boldsymbol{i}\mathbf{p} \\ -i\boldsymbol{k}E - \boldsymbol{i}\mathbf{p} \end{pmatrix} . \begin{pmatrix} -i\boldsymbol{k}E + \boldsymbol{i}\mathbf{p} \\ -i\boldsymbol{k}E - \boldsymbol{i}\mathbf{p} \\ i\boldsymbol{k}E + \boldsymbol{i}\mathbf{p} \\ i\boldsymbol{k}E - \boldsymbol{i}\mathbf{p} \end{pmatrix}$$

To construct a spin 0 boson structure we reverse the $\mathbf{p}$ signs in either fermion or antifermion, so that the components have the opposite spins but the same helicities:

$$\begin{pmatrix} i\boldsymbol{k}E + \boldsymbol{i}\mathbf{p} + \boldsymbol{j}m \\ i\boldsymbol{k}E - \boldsymbol{i}\mathbf{p} + \boldsymbol{j}m \\ -i\boldsymbol{k}E + \boldsymbol{i}\mathbf{p} + \boldsymbol{j}m \\ -i\boldsymbol{k}E - \boldsymbol{i}\mathbf{p} + \boldsymbol{j}m \end{pmatrix} . \begin{pmatrix} -i\boldsymbol{k}E - \boldsymbol{i}\mathbf{p} + \boldsymbol{j}m \\ -i\boldsymbol{k}E + \boldsymbol{i}\mathbf{p} + \boldsymbol{j}m \\ i\boldsymbol{k}E - \boldsymbol{i}\mathbf{p} + \boldsymbol{j}m \\ i\boldsymbol{k}E + \boldsymbol{i}\mathbf{p} + \boldsymbol{j}m \end{pmatrix}$$

The product is a nonzero scalar, but this time, if we reduce the mass to zero, we will zero the product as well.

$$\begin{pmatrix} i\boldsymbol{k}E + \boldsymbol{i}\mathbf{p} \\ i\boldsymbol{k}E - \boldsymbol{i}\mathbf{p} \\ -i\boldsymbol{k}E + \boldsymbol{i}\mathbf{p} \\ -i\boldsymbol{k}E - \boldsymbol{i}\mathbf{p} \end{pmatrix} . \begin{pmatrix} -i\boldsymbol{k}E - \boldsymbol{i}\mathbf{p} \\ -i\boldsymbol{k}E + \boldsymbol{i}\mathbf{p} \\ i\boldsymbol{k}E - \boldsymbol{i}\mathbf{p} \\ i\boldsymbol{k}E + \boldsymbol{i}\mathbf{p} \end{pmatrix} = 0$$

The implication of this is that a spin 0 boson, defined in this way, cannot be massless. So Goldstone bosons cannot exist, and the Higgs boson must have a mass. The mass works out, additionally, as will become apparent, as a measure of the degree of right-handedness in the fermion component and left-handedness in the antifermion component. In this context, we could say that a massless spin 0 boson must be made up of a left-handed fermion (weak-allowed) and a left-handed antifermion (weak-prohibited), $(\pm i\boldsymbol{k}E \pm \boldsymbol{i}\mathbf{p})(\mp i\boldsymbol{k}E \mp \boldsymbol{i}\mathbf{p})$, or a right-handed fermion (weak-prohibited) and a right-handed antifermion (weak-allowed), $(\pm i\boldsymbol{k}E \mp \boldsymbol{i}\mathbf{p})(\mp i\boldsymbol{k}E \pm \boldsymbol{i}\mathbf{p})$. To overcome the prohibited conditions, we must generate mass. Before the discovery of the Higgs boson, it was a matter of guesswork what its mass might be. On the basis of these structures, my own *final* guess was that the two weak-prohibited structures might each require half of the Higgs field, leading to a mass of 123 GeV. At 125 to 126 GeV, it looks closer to half of the mass of the boson structures (W^+, W^-, Z^0) which carry the weak interaction, and are directly generated from the field.

A third type of boson-like state can be formed by combining two fermions with opposite spins and opposite helicities:

$$\begin{pmatrix} i\boldsymbol{k}E + \boldsymbol{i}\mathbf{p} + \boldsymbol{j}m \\ i\boldsymbol{k}E - \boldsymbol{i}\mathbf{p} + \boldsymbol{j}m \\ -i\boldsymbol{k}E + \boldsymbol{i}\mathbf{p} + \boldsymbol{j}m \\ -i\boldsymbol{k}E - \boldsymbol{i}\mathbf{p} + \boldsymbol{j}m \end{pmatrix} \cdot \begin{pmatrix} i\boldsymbol{k}E - \boldsymbol{i}\mathbf{p} + \boldsymbol{j}m \\ i\boldsymbol{k}E + \boldsymbol{i}\mathbf{p} + \boldsymbol{j}m \\ -i\boldsymbol{k}E - \boldsymbol{i}\mathbf{p} + \boldsymbol{j}m \\ -i\boldsymbol{k}E + \boldsymbol{i}\mathbf{p} + \boldsymbol{j}m \end{pmatrix}$$

Again the product is a pure scalar. States of this kind can be imagined to occur in Cooper pairing in superconductors, in He^4 and Bose–Einstein condensates, in spin 0 nuclei, in the Jahn–Teller effect, the Aharonov–Bohm effect, the quantum Hall effect, and, in general, in states where there is a nonzero Berry phase to make fermions become single-valued in terms of spin.

In general, these will be spin 0 states, but a spin 1 fermion–fermion combination is known in the case of He^3. Here, the two components move with respect to each other with components of motion in opposite directions, presumably in some kind of harmonic oscillator fashion, meaning that they could have the same spin states but opposite helicities.

In the case of the fermion–fermion spin 0 state, we can make a prediction which can be tested by experiment. Because the formation of this state necessarily requires intrinsically massive components, even in those cases where it assumes nonzero effective mass through a Fermi velocity less than c, time reversal symmetry (the one applicable to the transition) must be broken in the weak formation or decay of such states. The most likely opportunity of observing such a process might be in one of the physical manifestations of the nonzero Berry phase, say the quantum Hall effect, in some special type of condensed matter such as graphene. Here, the conduction electrons have zero effective mass and a Hamiltonian that can be written in the form $\pm v_F \boldsymbol{i}(\mathbf{i}p_x + \mathbf{j}p_y)$, where v_F is the Fermi velocity. We can imagine creating a boson-like state with single-valued spin by the quantum Hall effect, Aharonov–Bohm effect, or Bose–Einstein condensation, and then observing, perhaps through a change in the Fermi velocity during its decay, the violation of both P and $CP = T$ symmetries.

Bosons, we have said, cannot be seen as independent of fermions. If a fermion could combine with its own vacuum, it would annihilate automatically, but this is, of course, impossible because this vacuum represents the entire universe outside of the fermion; however, we can imagine it combining with a *component* of vacuum, or the fermionic state equivalent to this. The various bosonic states may be seen as simultaneous realisations of the two vector spaces involved in the creation of the fermionic state, though at the expense of making only one component of the vacuum space well defined, just as only one component of angular momentum is well defined in real space.

Spin 1 bosons are involved in all local interactions, but they occur in a particular way in weak interactions. Here, fermions and antifermions are annihilated while bosons are created, or bosons are annihilated while fermions and antifermions are created, and, frequently, both processes (or equivalent) occur. We recognise the creation and annihilation of states as the action of a harmonic oscillator. It can also be thought of as the creation and annihilation of the weak source (or weak 'charge'). A very important difference between fermions and bosons is that fermions are sources for weak interactions, while bosons are not. The W and Z bosons are carriers of the weak force, but are not sources of it. Bosons, considered to be created at fermion–antifermion vertices, are the products of weak interactions. Even in examples such as electron–positron collisions, where the predominant interaction is electric at low energies, there is an amplitude for a weak interaction.

6.2 Baryons

We have postulated that fermions are created as singularities through a dual space structure. Baryons complicate this structure by introducing an explicit 3-dimensionality into the real space part of the structure. They are an expression of the fact that singularities are incompatible with a pure 3-dimensional space and are only possible where we have a dual space. The strong interaction begins in a combination state, which reflects the *vector* nature of the $\mathbf{p}$ term in the nilpotent wavefunction. Essentially, the 'quarks' in a baryon are like components of a vector; they can no more be separated than the directions in space. Effectively, the vector aspect of the strong charge requires a source term and corresponding vacuum with three components. Though we clearly cannot combine three components in the form:

$$(i\boldsymbol{k}E \pm \boldsymbol{i}\mathbf{p} + \boldsymbol{j}m)(i\boldsymbol{k}E \pm \boldsymbol{i}\mathbf{p} + \boldsymbol{j}m)(i\boldsymbol{k}E \pm \boldsymbol{i}\mathbf{p} + \boldsymbol{j}m)$$

as this will automatically reduce to zero, we can imagine a three-component structure in which the vector nature of $\mathbf{p}$ plays an explicit role:

$$(i\boldsymbol{k}E \pm \boldsymbol{i}\mathbf{i}p_x + \boldsymbol{j}m)(i\boldsymbol{k}E \pm \boldsymbol{i}\mathbf{j}p_y + \boldsymbol{j}m)(i\boldsymbol{k}E \pm \boldsymbol{i}\mathbf{k}p_z + \boldsymbol{j}m)$$

This has nilpotent solutions when $\mathbf{p} = \pm \boldsymbol{i}\mathbf{i}p_x, \mathbf{p} = \pm \boldsymbol{i}\mathbf{j}p_y$, or $\mathbf{p} = \pm \boldsymbol{i}\mathbf{k}p_z$, or when the momentum is directed entirely along the x, y, or z axes, in either direction, though these, of course, are arbitrarily defined. For convenience, we have written only the first term of the 4-component spinors, but we have retained the two spin states, as these will be needed explicitly.

The complete wavefunction will, in effect, contain information from the equivalent of six allowed independent nonlocally gauge invariant phases, all existing simultaneously and subject to continual transitions at a constant rate:

$$\begin{array}{ll}
(i\boldsymbol{k}E + \boldsymbol{i}\mathbf{i}p_x + \boldsymbol{j}m)(i\boldsymbol{k}E + 0 + \boldsymbol{j}m)(i\boldsymbol{k}E + 0 + \boldsymbol{j}m) & +RGB \\
(i\boldsymbol{k}E - \boldsymbol{i}\mathbf{i}p_x + \boldsymbol{j}m)(i\boldsymbol{k}E - 0 + \boldsymbol{j}m)(i\boldsymbol{k}E - 0 + \boldsymbol{j}m) & -RBG \\
(i\boldsymbol{k}E + 0 + \boldsymbol{j}m)(i\boldsymbol{k}E + \boldsymbol{i}\mathbf{j}p_y + \boldsymbol{j}m)(i\boldsymbol{k}E + 0 + \boldsymbol{j}m) & +BRG \\
(i\boldsymbol{k}E - 0 + \boldsymbol{j}m)(i\boldsymbol{k}E - \boldsymbol{i}\mathbf{j}p_y + \boldsymbol{j}m)(i\boldsymbol{k}E - 0 + \boldsymbol{j}m) & -GRB \\
(i\boldsymbol{k}E + 0 + \boldsymbol{j}m)(i\boldsymbol{k}E + 0 + \boldsymbol{j}m)(i\boldsymbol{k}E + \boldsymbol{i}\mathbf{k}p_z + \boldsymbol{j}m) & +GBR \\
(i\boldsymbol{k}E - 0 + \boldsymbol{j}m)(i\boldsymbol{k}E - 0 + \boldsymbol{j}m)(i\boldsymbol{k}E - \boldsymbol{i}\mathbf{k}p_z + \boldsymbol{j}m) & -BGR
\end{array}$$

Any other phases can be written as a superposition of these. Using the appropriate normalization, these reduce to

$$\begin{array}{ll}
(i\boldsymbol{k}E + \boldsymbol{i}\mathbf{i}p_x + \boldsymbol{j}m) & +RGB \\
(i\boldsymbol{k}E - \boldsymbol{i}\mathbf{i}p_x + \boldsymbol{j}m) & -RBG \\
(i\boldsymbol{k}E - \boldsymbol{i}\mathbf{j}p_y + \boldsymbol{j}m) & +BRG
\end{array}$$

$$\begin{array}{ll} (i\boldsymbol{k}E + \boldsymbol{i}\mathbf{j}p_y + \boldsymbol{j}m) & -GRB \\ (i\boldsymbol{k}E + \boldsymbol{i}\mathbf{k}p_z + \boldsymbol{j}m) & +GBR \\ (i\boldsymbol{k}E - \boldsymbol{i}\mathbf{k}p_z + \boldsymbol{j}m) & -BGR \end{array}$$

with the third and fourth changing, very significantly, the sign of the **p** component. Because of this, there has to be a maximal superposition of left- and right-handed components, thus explaining the zero observed chirality in the interaction.

The group structure required to maintain these phases is an $SU(3)$ structure with eight generators and a wavefunction, exactly as in the conventional model using coloured quarks,

$$\psi \sim (BGR - BRG + GRB - GBR + RBG - RGB).$$

'Colour' transitions in the 3-component structures are produced either by an exchange of the components of **p** between the individual quarks or baryon components, or by a relative switching of the component positions independently of any real distance between the components. No direction can be privileged, so the transition must be gauge invariant, and the mediators must be massless, exactly as with the eight massless gluons of the gluon structure. Here, six gluons can be constructed from:

$$\begin{array}{ll} (i\boldsymbol{k}E + \boldsymbol{i}\mathbf{i}p_x)(-i\boldsymbol{k}E + \boldsymbol{i}\mathbf{j}p_y) & (i\boldsymbol{k}E + \boldsymbol{i}\mathbf{j}p_y)(-i\boldsymbol{k}E + \boldsymbol{i}\mathbf{i}p_x) \\ (i\boldsymbol{k}E + \boldsymbol{i}\mathbf{j}p_y)(-i\boldsymbol{k}E + \boldsymbol{i}\mathbf{k}p_z) & (i\boldsymbol{k}E + \boldsymbol{i}\mathbf{k}p_z)(-i\boldsymbol{k}E + \boldsymbol{i}\mathbf{j}p_y) \\ (i\boldsymbol{k}E + \boldsymbol{i}\mathbf{k}p_z)(-i\boldsymbol{k}E + \boldsymbol{i}\mathbf{i}p_x) & (i\boldsymbol{k}E + \boldsymbol{i}\mathbf{i}p_x)(-i\boldsymbol{k}E + \boldsymbol{i}\mathbf{k}p_z) \end{array}$$

and two from combinations of

$$\begin{array}{ll} (i\boldsymbol{k}E + \boldsymbol{i}\mathbf{i}p_x)(-i\boldsymbol{k}E + \boldsymbol{i}\mathbf{i}p_x) & (i\boldsymbol{k}E + \boldsymbol{i}\mathbf{j}p_y)(-i\boldsymbol{k}E + \boldsymbol{i}\mathbf{j}p_y) \\ (i\boldsymbol{k}E + \boldsymbol{i}\mathbf{k}p_z)(-i\boldsymbol{k}E + \boldsymbol{i}\mathbf{k}p_z) & \end{array}$$

Alternatively, we can switch the signs, producing an equivalent set:

$$\begin{array}{ll} (i\boldsymbol{k}E - \boldsymbol{i}\mathbf{i}p_x)(-i\boldsymbol{k}E - \boldsymbol{i}\mathbf{j}p_y) & (i\boldsymbol{k}E - \boldsymbol{i}\mathbf{j}p_y)(-i\boldsymbol{k}E - \boldsymbol{i}\mathbf{i}p_x) \\ (i\boldsymbol{k}E - \boldsymbol{i}\mathbf{j}p_y)(-i\boldsymbol{k}E - \boldsymbol{i}\mathbf{k}p_z) & (i\boldsymbol{k}E - \boldsymbol{i}\mathbf{k}p_z)(-i\boldsymbol{k}E - \boldsymbol{i}\mathbf{j}p_y) \\ (i\boldsymbol{k}E - \boldsymbol{i}\mathbf{k}p_z)(-i\boldsymbol{k}E - \boldsymbol{i}\mathbf{i}p_x) & (i\boldsymbol{k}E - \boldsymbol{i}\mathbf{i}p_x)(-i\boldsymbol{k}E - \boldsymbol{i}\mathbf{k}p_z) \end{array}$$

with two combinations from

$$\begin{array}{ll} (i\boldsymbol{k}E - \boldsymbol{i}\mathbf{i}p_x)(-i\boldsymbol{k}E - \boldsymbol{i}\mathbf{i}p_x) & (i\boldsymbol{k}E - \boldsymbol{i}\mathbf{j}p_y)(-i\boldsymbol{k}E - \boldsymbol{i}\mathbf{j}p_y) \\ (i\boldsymbol{k}E - \boldsymbol{i}\mathbf{k}p_z)(-i\boldsymbol{k}E - \boldsymbol{i}\mathbf{k}p_z) & \end{array}$$

where, as with the baryons, only the lead term is shown for each 4-component spinor.

A representation such as the 3-component baryon above, showing only one 'quark' active at any time in contributing to the angular momentum operator, seems to indicate why only 1/3 of baryon spin has been found to be due to the valence quarks. The rest of the spin then becomes a 'vacuum' contribution split approximately 3 to 1 in favour of the gluons over the sea quarks, the gluons thus taking half of the overall total. The simultaneous existence of all phases further means that *individual* quarks, and such identifying characteristics as electric charges, are not identifiable by their spatial positions (unlike, say, the proton and electron constituting a hydrogen atom), thus explaining, for example, why the neutron has no electric dipole moment. As we established in Chapter 4, just as $U(1)$ ensures that spherical symmetry of a point source requires the rotation to be performed independently of the length of the radius vector, $SU(3)$ requires the rotation to be independent of the coordinate system used. In terms of Noether's theorem, while $U(1)$ conserves the magnitude of angular momentum, $SU(3)$ conserves the direction.

The structures derived here produce insights into at least two fundamental physical problems. The first is the mass-gap problem for baryons, which is one of the Clay Institute's Millennium Prize Problems. We are confronted with the fact that baryons have nonzero mass and yet this mass is thought to be produced by the action of massless gluons. And, although the Higgs mechanism appears to be the main process by which mass is delivered to fermions, the gluon exchange is generally considered to be a non-Higgs process. In fact, the 3-component structures clearly require the simultaneous existence of two states of helicity for the symmetry to remain unbroken (because the placing of the middle component ensures that there is a sign switch in $\mathbf{p}$), and this can only be possible if the baryon has nonzero mass.

In addition, this process is the signature of the Higgs mechanism, and so, contrary to much current supposition, the generation of the masses of baryons follows exactly the same process as that of all other fermions. However, this does not contradict the fact, established by much calculation using QCD, that the bulk of the mass of a baryon is due to the exchange of massless gluons, as the exchange of gluons structured as above will necessarily lead to a sign change in the $\mathbf{p}$ operator, and hence of helicity, the exact mechanism which is responsible for the production of all known particle masses. In fact, the same will be true of all fermions involved in spin 1 boson exchange, and so, once again, all fermions must have nonzero masses.

The second problem is the specific nature and mechanism of the strong interaction between quarks. Again, we see that a solution is suggested by the exact structure of the nilpotent operator. Here, we already know that there must be a Coulomb component or inverse linear potential ($\propto 1/r$) to accommodate spherical symmetry. This has a known physical manifestation in the one-gluon exchange. But there is also at least one other component, which is responsible for quark confinement, for infrared slavery and for asymptotic freedom, and a linear potential ($\propto r$) has long been hypothesised and used in calculations. Here, we see that an exchange of $\mathbf{p}$ components at a constant rate would, in principle, require a constant rate of change of momentum, which is the signature of a linear potential.

6.3 Partitioning the vacuum

The nilpotent formalism defines a continuous vacuum $-(\pm \boldsymbol{ik}E \pm \boldsymbol{i}\mathbf{p} + \boldsymbol{j}m)$ to each fermionic state $(\pm \boldsymbol{ik}E \pm \boldsymbol{i}\mathbf{p} + \boldsymbol{j}m)$, and this vacuum expresses the nonlocal aspect of the state. However, the use of the operators $\boldsymbol{k}$, $\boldsymbol{i}$, $\boldsymbol{j}$ suggests that we can partition this state into discrete components with a dimensional structure. In fact, this is where the idempotents become relevant. If we postmultiply $(\pm \boldsymbol{ik}E \pm \boldsymbol{i}\mathbf{p} + \boldsymbol{j}m)$ by the idempotent $\boldsymbol{k}(\pm \boldsymbol{ik}E \pm \boldsymbol{i}\mathbf{p} + \boldsymbol{j}m)$ any number of times, the only change is to introduce a scalar multiple, which can be normalized away.

$$(\pm \boldsymbol{ik}E \pm \boldsymbol{i}\mathbf{p} + \boldsymbol{j}m)\boldsymbol{k}(\pm \boldsymbol{ik}E \pm \boldsymbol{i}\mathbf{p} + \boldsymbol{j}m)\boldsymbol{k}(\pm \boldsymbol{ik}E \pm \boldsymbol{i}\mathbf{p} + \boldsymbol{j}m)\cdots$$
$$\rightarrow (\pm \boldsymbol{ik}E \pm \boldsymbol{i}\mathbf{p} + \boldsymbol{j}m)$$

The idempotent acts as a vacuum operator. The same applies to postmultiplication by $\boldsymbol{i}(\pm \boldsymbol{ik}E \pm \boldsymbol{i}\mathbf{p} + \boldsymbol{j}m)$ or $\boldsymbol{j}(\pm \boldsymbol{ik}E \pm \boldsymbol{i}\mathbf{p} + \boldsymbol{j}m)$, except that the former also produces a unit vector which disappears on every alternate postmultiplication and has no effect on the nilpotent $(\pm \boldsymbol{ik}E \pm \boldsymbol{i}\mathbf{p} + \boldsymbol{j}m)$. Now, the multiplication by $\boldsymbol{k}(\pm \boldsymbol{ik}E \pm \boldsymbol{i}\mathbf{p} + \boldsymbol{j}m)$ is also equivalent to applying a time-reversal transformation to every even $(\pm \boldsymbol{ik}E \pm \boldsymbol{i}\mathbf{p} + \boldsymbol{j}m)$. So we have

$$(\pm \boldsymbol{ik}E \pm \boldsymbol{i}\mathbf{p} + \boldsymbol{j}m)(\mp \boldsymbol{ik}E \pm \boldsymbol{i}\mathbf{p} + \boldsymbol{j}m)(\pm \boldsymbol{ik}E \pm \boldsymbol{i}\mathbf{p} + \boldsymbol{j}m)\cdots$$
$$\rightarrow (\pm \boldsymbol{ik}E \pm \boldsymbol{i}\mathbf{p} + \boldsymbol{j}m)$$

with every alternate state becoming an antifermion, which combines with the original fermion state to become a spin 1 boson $(\pm i\boldsymbol{k}E \pm \boldsymbol{i}\mathbf{p} + \boldsymbol{j}m)(\mp i\boldsymbol{k}E \pm \boldsymbol{i}\mathbf{p} + \boldsymbol{j}m)$.

Similar results are obtained with $\boldsymbol{i}(\pm i\boldsymbol{k}E \pm \boldsymbol{i}\mathbf{p} + \boldsymbol{j}m)$ or $\boldsymbol{j}(\pm i\boldsymbol{k}E \pm \boldsymbol{i}\mathbf{p} + \boldsymbol{j}m)$, which produce spin 0 bosons and bosonic paired fermions, via a parity transformation and a charge conjugation. It looks like that, from an initial fermionic state, we can generate either three vacuum reflections, via respective T, P and C transformations, representing antifermion with the same spin, fermion with opposite spin, and antifermion with opposite spin, or combined particle–vacuum states which have the respective structures of spin 1 bosons, spin 0 bosons, or boson-like paired fermion (PF) combinations of the same kind as constitute Cooper pairs and the elements of Bose–Einstein condensates. Using just the lead terms of the nilpotents, and assuming that we can complete the spinor structures using the 3 conventional sign variations, we could represent these as:

$$\begin{aligned}
&(i\boldsymbol{k}E + \boldsymbol{i}\mathbf{p} + \boldsymbol{j}m)\boldsymbol{k}(i\boldsymbol{k}E + \boldsymbol{i}\mathbf{p} + \boldsymbol{j}m) \\
&\quad \times \boldsymbol{k}(i\boldsymbol{k}E + \boldsymbol{i}\mathbf{p} + \boldsymbol{j}m)\boldsymbol{k}(i\boldsymbol{k}E + \boldsymbol{i}\mathbf{p} + \boldsymbol{j}m)\ldots \qquad T \\
&(i\boldsymbol{k}E + \boldsymbol{i}\mathbf{p} + \boldsymbol{j}m)(-i\boldsymbol{k}E + \boldsymbol{i}\mathbf{p} + \boldsymbol{j}m) \\
&\quad \times (i\boldsymbol{k}E + \boldsymbol{i}\mathbf{p} + \boldsymbol{j}m)(-i\boldsymbol{k}E + \boldsymbol{i}\mathbf{p} + \boldsymbol{j}m)\ldots \qquad \text{spin 1} \\
&(i\boldsymbol{k}E + \boldsymbol{i}\mathbf{p} + \boldsymbol{j}m)\boldsymbol{j}(i\boldsymbol{k}E + \boldsymbol{i}\mathbf{p} + \boldsymbol{j}m) \\
&\quad \times \boldsymbol{j}(i\boldsymbol{k}E + \boldsymbol{i}\mathbf{p} + \boldsymbol{j}m)\boldsymbol{j}(i\boldsymbol{k}E + \boldsymbol{i}\mathbf{p} + \boldsymbol{j}m)\ldots \qquad C \\
&(i\boldsymbol{k}E + \boldsymbol{i}\mathbf{p} + \boldsymbol{j}m)(-i\boldsymbol{k}E - \boldsymbol{i}\mathbf{p} + \boldsymbol{j}m) \\
&\quad \times (i\boldsymbol{k}E + \boldsymbol{i}\mathbf{p} + \boldsymbol{j}m)(-i\boldsymbol{k}E - \boldsymbol{i}\mathbf{p} + \boldsymbol{j}m)\ldots \qquad \text{spin 0} \\
&(i\boldsymbol{k}E + \boldsymbol{i}\mathbf{p} + \boldsymbol{j}m)\boldsymbol{i}(i\boldsymbol{k}E + \boldsymbol{i}\mathbf{p} + \boldsymbol{j}m) \\
&\quad \times \boldsymbol{i}(i\boldsymbol{k}E + \boldsymbol{i}\mathbf{p} + \boldsymbol{j}m)\boldsymbol{i}(i\boldsymbol{k}E + \boldsymbol{i}\mathbf{p} + \boldsymbol{j}m)\ldots \qquad P \\
&(i\boldsymbol{k}E + \boldsymbol{i}\mathbf{p} + \boldsymbol{j}m)(i\boldsymbol{k}E - \boldsymbol{i}\mathbf{p} + \boldsymbol{j}m) \\
&\quad \times (i\boldsymbol{k}E + \boldsymbol{i}\mathbf{p} + \boldsymbol{j}m)(i\boldsymbol{k}E - \boldsymbol{i}\mathbf{p} + \boldsymbol{j}m)\ldots \qquad \text{PF}
\end{aligned}$$

These processes indicate that repeated postmultiplication of a fermionic operator by any of the discrete idempotent vacuum operators creates an alternate series of antifermion and fermion vacuum states, or, equivalently, an alternate series of bosonic and fermionic states without changing the character of the real particle state. A fermion produces a bosonic state by combining with its own vacuum image, and the two states form a *supersymmetric* partnership. Nilpotent operators are thus intrinsically supersymmetric, with supersymmetry operators typically of the form:

$$\text{Boson to fermion:} \quad Q = (\pm i\boldsymbol{k}E \pm \boldsymbol{i}\mathbf{p} + \boldsymbol{j}m)$$
$$\text{Fermion to boson:} \quad Q^\dagger = (\mp i\boldsymbol{k}E \pm \boldsymbol{i}\mathbf{p} + \boldsymbol{j}m)$$

A fermion converts to a boson by multiplication by an antifermionic operator $Q^\dagger$; a boson converts to a fermion by multiplication by a fermionic operator Q, and we can represent the sequence $(i\boldsymbol{k}E + \boldsymbol{i}\mathbf{p} + \boldsymbol{j}m)\boldsymbol{k}(i\boldsymbol{k}E + \boldsymbol{i}\mathbf{p} + \boldsymbol{j}m) \ldots$ by the supersymmetric

$$QQ^\dagger QQ^\dagger QQ^\dagger QQ^\dagger Q \ldots$$

We will see in the next chapter that we can interpret this as the series of boson and fermion loops of the same energy and momentum, required by the exact supersymmetry which would eliminate the need for renormalization, and remove the hierarchy problem altogether. Fermions and bosons (with the same values E, $\mathbf{p}$ and m) then become their own supersymmetric partners through the creation of vacuum states, making the hypothesis of a set of real supersymmetric particles to solve the hierarchy problem potentially superfluous.

The identification of $\boldsymbol{i}(i\boldsymbol{k}E + \boldsymbol{i}\mathbf{p} + \boldsymbol{j}m)$, $\boldsymbol{k}(i\boldsymbol{k}E + \boldsymbol{i}\mathbf{p} + \boldsymbol{j}m)$ and $\boldsymbol{j}(i\boldsymbol{k}E + \boldsymbol{i}\mathbf{p} + \boldsymbol{j}m)$ as vacuum operators and $(i\boldsymbol{k}E - \boldsymbol{i}\mathbf{p} + \boldsymbol{j}m), (-i\boldsymbol{k}E + \boldsymbol{i}\mathbf{p} + \boldsymbol{j}m)$ and $(-i\boldsymbol{k}E - \boldsymbol{i}\mathbf{p} + \boldsymbol{j}m)$ as their respective vacuum 'reflections' at interfaces provided by P, T and C transformations suggests a new insight into the meaning of the Dirac 4-spinor. We can now interpret the three terms other than the lead term *in the spinor* as the vacuum 'reflections' that are created with the particle. We can regard the existence of three vacuum operators as a result of a partitioning of the vacuum stemming from quantization and as a consequence of the 3-part structure observed in the nilpotent fermionic state, while the *zitterbewegung* can be taken as an indication that the vacuum is active in defining the fermionic state.

The four components of the spinor cancel exactly when the operator or amplitude is written in nilpotent mode. This is even more apparent if the components are represented as operators using discrete calculus when the cancellation is a zero algebraic sum. (In idempotent mode, the summed amplitudes would normalize to 1.) Though annihilation and creation operators are 'black box' devices in standard quantum field theory, here, like the wavefunctions, they have exact and explicit mathematical and physical representations. We can thus represent the four components of the nilpotent spinor as creation operators for

fermion spin up	$(i\boldsymbol{k}E + \mathbf{ip} + \boldsymbol{j}m)$
fermion spin down	$(i\boldsymbol{k}E - \mathbf{ip} + \boldsymbol{j}m)$
antifermion spin down	$(-i\boldsymbol{k}E + \mathbf{ip} + \boldsymbol{j}m)$
antifermion spin up	$(-i\boldsymbol{k}E - \mathbf{ip} + \boldsymbol{j}m)$

or annihilation operators for

antifermion spin down	$(i\boldsymbol{k}E + \mathbf{ip} + \boldsymbol{j}m)$
antifermion spin up	$(i\boldsymbol{k}E - \mathbf{ip} + \boldsymbol{j}m)$
fermion spin up	$(-i\boldsymbol{k}E + \mathbf{ip} + \boldsymbol{j}m)$
fermion spin down	$(-i\boldsymbol{k}E - \mathbf{ip} + \boldsymbol{j}m)$

They can equally well be regarded as two operators for creation and two for annihilation, for example,

fermion spin up creation	$(i\boldsymbol{k}E + \mathbf{ip} + \boldsymbol{j}m)$
fermion spin down creation	$(i\boldsymbol{k}E - \mathbf{ip} + \boldsymbol{j}m)$
fermion spin up annihilation	$(-i\boldsymbol{k}E + \mathbf{ip} + \boldsymbol{j}m)$
fermion spin down annihilation	$(-i\boldsymbol{k}E - \mathbf{ip} + \boldsymbol{j}m)$

In all cases, the cancellation is exact both physically and algebraically (when we use the discrete operators and leave out the passive mass component). It is interesting that the cancellation requires *four* components, rather than two, for, while the transitions:

$$(i\boldsymbol{k}E + \mathbf{ip} + \boldsymbol{j}m) \to (i\boldsymbol{k}E - \mathbf{ip} + \boldsymbol{j}m)$$

and

$$(i\boldsymbol{k}E + \mathbf{ip} + \boldsymbol{j}m) \to (-i\boldsymbol{k}E + \mathbf{ip} + \boldsymbol{j}m)$$

can occur through spin 1 boson and spin 0 paired fermion exchange, and the active space and time components, there is no process in nature for the *direct* transition:

$$(i\boldsymbol{k}E + \mathbf{ip} + \boldsymbol{j}m) \to (-i\boldsymbol{k}E - \mathbf{ip} + \boldsymbol{j}m)$$

with no active component as agent. In this context, it might be worth noting that the spin 0 fermion–fermion state

$$(i\boldsymbol{k}E + \mathbf{ip} + \boldsymbol{j}m)(i\boldsymbol{k}E - \mathbf{ip} + \boldsymbol{j}m)$$

is such as would be required in a pure weak transition from $-i\boldsymbol{k}E$ to $+i\boldsymbol{k}E$, or its inverse.

A special form of the generation of bosons by the process of fermion annihilation and creation occurs with *zitterbewegung*, where there is continually switching between the four fermionic states, leading to the annihilation of one state and the creation of another, so fulfilling the conditions for a weak interaction. This is, of course, a vacuum process, and the 'supersymmetric' bosons it creates can be thought of as the result of every fermion interacting weakly with its own vacuum reflection. In a sense, every weak source (or fermion) acts as a weak dipole, the second 'pole' being the vacuum reflection, which, in a gauge invariant system, exists simultaneously with the real 'particle' state. Of the two allowed 'direct' transitions, the one that switches between a fermion in real space and an antifermion in vacuum space, with the opposite helicity (a result of the pseudoscalar nature of the coefficient of $\boldsymbol{ik}E$), requires a spin 0 bosonic state, and so is a natural mass generator. The other direct transition is between spin up and spin down, which both exist in the real state of a fermion with nonzero mass.

The weak interaction is clearly related to the nature of the pseudoscalar iE operator, whose sign uniquely determines the helicity of a weakly interacting particle, or more specifically its weakly interacting component. It also has a unique feature in that its fermionic source cannot be separated from its vacuum partner. A fermion or antifermion cannot be created or annihilated, even with an antifermionic or fermionic partner, unless its vacuum is simultaneously annihilated or created. In this sense, the weak source has a manifestly dipolar nature, whose immediate manifestation is the fermion's $\frac{1}{2}$-integral spin. It is the most direct evidence we have of the duality of the vector space structure which underlies quantum physics.

Now, we know that the three vacuum coefficients $\boldsymbol{k}$, $\boldsymbol{i}$, $\boldsymbol{j}$ originate in (or are responsible for) the concept of discrete (point-like) charge. However, the operators, $\boldsymbol{k}$, $\boldsymbol{i}$ and $\boldsymbol{j}$, as we are using them here, perform another function of weak, strong and electric 'charges' or sources, in acting to partition the *continuous* vacuum represented by $-(\boldsymbol{ik}E + \boldsymbol{i}\mathbf{p} + \boldsymbol{j}m)$, and responsible for zero-point energy, into discrete components, whose special characteristics are determined by the respective pseudoscalar, vector and scalar natures of their associated terms iE, $\mathbf{p}$ and m.

In this way, they become related to the 'real' weak, strong and electric localised charges, though they are delocalised. We can describe the partitions as strong, weak and electric 'vacua', and assign to them particular roles within existing physics:

$\boldsymbol{k}(i\boldsymbol{k}E + \boldsymbol{i}\mathbf{p} + \boldsymbol{j}m)$	weak vacuum	fermion creation
$\boldsymbol{i}(i\boldsymbol{k}E + \boldsymbol{i}\mathbf{p} + \boldsymbol{j}m)$	strong vacuum	gluon plasma
$\boldsymbol{j}(i\boldsymbol{k}E + \boldsymbol{i}\mathbf{p} + \boldsymbol{j}m)$	electric vacuum	isospin/hypercharge

These three vacua retain the characteristics of the generating charge structures, respectively pseudoscalar, vector and scalar, which explain also the special characteristics and group structures of the forces with which they are associated. It is the vector characteristic of the strong vacuum that makes baryonic structure possible, and it is the pseudoscalar characteristic of the weak vacuum that makes the link between particle structure and vacuum possible at all. The 'electric vacuum' — empty or filled — can be seen as responsible for the transition between weak isospin up and down states (see Section 6.7).

6.4 Local and nonlocal

Many people think that nonlocality is a problem for quantum mechanics, but, in fact, it is a necessary component of any physics whose operations are truly universal. It is also an essential step towards understanding how *local* interactions actually work. Locality and nonlocality are not opposed concepts. They are two aspects of the description of any process in nature. They are parts of a dual system in which each aspect determines the behaviour of the other. The reason for the duality is simple. If we are describing the behaviour of a point source, we can start by specifying what the point is, or by what it isn't, i.e. everything else. The first is the local description, the second the nonlocal. It also represents the duality at the heart of quantum physics: real space requires its dual in vacuum space.

The holistic nature of physics means that the local cannot, in fact, be separated from the nonlocal. Even the terms can be misleading, because 'local' refers to the entire universe as much as nonlocal does. The nilpotent version of quantum mechanics shows that there is no such thing as an isolated system, and so a complete local description, which originates in the individual particle, will still require knowledge of the contents and disposition of the whole universe. In principle, the difference between local and nonlocal is not in the phenomena they describe, but in the method of description, essentially whether we use an iterative or recursive computational paradigm. Local interactions are determined by the collective nonlocal effect of the rest of the universe.

When we write down an operator or amplitude in the form $(\pm i\boldsymbol{k}E \pm \boldsymbol{i}\mathbf{p} + \boldsymbol{j}m)$, the brackets may suggest that we have created a closed system, but in fact the E and $\mathbf{p}$ terms may contain an unlimited number of potentials. We have created a system but it is open. Closure or energy conservation is only maintained over the entire universe, and requires the second law of thermodynamics as well as the first. So, though the bracket may define locality, locality does not imply a closed system. The creation of the fermionic state is the creation of a local region in phase space to which everything else becomes nonlocal; the creation of the two regions is simultaneous. Any subsequent change inside the bracket (a rewriting of the structures of E and $\mathbf{p}$) also affects everything else outside it, and vice versa.

We can now show that the exact characteristics of the different local interactions (electric, strong and weak) are completely determined by the nonlocal vacuum structures with which they are associated. Beginning with the nonlocal characteristics that emerge from the way that the fermionic state is structured as a nilpotent operator, we can derive the *exact form* of the potentials that would produce the same result by a *local* process. We can then find analytic nilpotent solutions in all these cases which provide exact matches to the results found for the electric, strong and weak interactions from experiment, and from their descriptions in terms of the $U(1)$, $SU(3)$ and $SU(2)$ symmetry groups. We will also show that these solutions are unique. No others are permitted within the nilpotent algebra. In other words, we can present a completely integrated description of local interactions and nonlocal vacuum structures based on nilpotent quantum mechanics and its unique algebraic structure. The complete set of structures depends entirely on the algebraic symmetry-breaking that emerges from the creation of point sources from two vector spaces, which are dual to each other. In effect, we propose to show in a completely analytical form that the two-space algebra leads to the entire basis of the interactions required by the Standard Model.

We have three interactions to consider, and we have three nonlocal aspects of the nilpotent wavefunction from which we believe they can be derived. By nonlocal, we mean any operation that occurs outside a nilpotent bracket. The two principal examples are superposition, where brackets are added, and combination, where brackets are multiplied. Nonlocal operations are instantaneous and occur across the entire universe. Local operations are Lorentz-invariant and are limited by the maximum speed c. They are determined entirely by what happens inside the bracket. However, nonlocal operations have local manifestations and vice versa. Decoherence of a quantum system (the so-called 'collapse of the wavefunction') is a clear example of the reverse transformation, as it uses local potentials to remove the nonlocal quantum coherence. This is how we propose to solve the 'measurement problem'.

The first nonlocal operation is nilpotency itself, or Pauli exclusion, with each fermion instantaneously ensuring that it has a different phase and amplitude to any other. We have seen that this is the result of the creation of a fixed point source with spherical symmetry, that it demands a Coulomb term (potential energy $\propto 1/r$) in the energy operator, and that it is required in all possible local interactions. It is determined solely by scale, as it is the scalar values of E, p and m that fix nilpotency, and so is related to the sources associated with $\boldsymbol{k}$, $\boldsymbol{i}$ and $\boldsymbol{j}$. The addition of the inverse linear potential to the E term changes the nonlocal $U(1)$ (nilpotent or Pauli exclusion) condition to a local one, since the potential is added *inside* the bracket; and this becomes the pattern for all the interactions.

The second is the combination state formed between three-component wavefunctions, expressing the vector nature of the operator $\mathbf{p}$, and we have seen that this requires an additional linear potential energy ($\propto r$) to add to the Coulomb term in its *local* formulation. Because of its origin in the $\mathbf{p}$ term, it originates in the source associated with $\boldsymbol{i}$. The third is the superposition state created by the four terms in the fermion spinor, which manifests itself as two fermion terms in real space (with $i\boldsymbol{k}E$) and two antifermion terms in vacuum space (with $-i\boldsymbol{k}E$), and which in turn result from the fact that the energy term iE is a pseudoscalar with dual $+$ and $-$ values. Locally, this requires the source to act as a dipole (at least — it may be more complicated) as well as monopole, which requires an additional potential energy term proportional to $1/r^2$, $1/r^3$, or some other $1/r^n$. This clearly originates in the source connected to $\boldsymbol{k}$. Of course, these interactions also have manifestations which extend beyond the individual fermion, but these will be of exactly the same character. In fact, all weak interactions operate according to the same mechanism, and the mediator is a spin 1 boson $(\pm i\boldsymbol{k}E \pm \boldsymbol{i}\mathbf{p} + \boldsymbol{j}m)\boldsymbol{k}(\pm i\boldsymbol{k}E \pm \boldsymbol{i}\mathbf{p} + \boldsymbol{j}m)\boldsymbol{k}$, equivalent to the state produced by a fermion acting on a 'weak' vacuum.

It would appear, then, that the nonlocal characteristics also require local manifestations appearing in the E or $\mathbf{p}$ terms of the nilpotent wavefunction. Here, we can restrict them to E by choice of frame. Having made no prior assumption about the nature of these interactions, we can assign the name of 'strong' to the one whose source is associated with $\boldsymbol{i}\mathbf{p}$, and the name of 'weak' to the one whose source is associated with $i\boldsymbol{k}E$, leaving the name 'electric' for the one whose source is associated with $\boldsymbol{j}m$, and has a purely Coulomb manifestation. We will now proceed to find the nilpotent structure corresponding to these three cases.

6.5 The Coulomb (electric) interaction

We have already produced the nilpotent operator with the required Coulomb term, we will find that it can be solved, using the known procedures but eliminating many unnecessary ones, in only six lines of calculation. This may seem surprising in view of the complexities of calculations already available, but calculations are notably easier and more efficient in the nilpotent structure than in alternative formalisms, largely because *dual* information, concerning both fermion and vacuum, is available, as is completely new *physical* information. So we begin with:

$$\left(\pm i\boldsymbol{k} \left(E - \frac{A}{r} \right) \mp i\boldsymbol{i} \left(\frac{\partial}{\partial r} + \frac{1}{r} \pm i \frac{j + \frac{1}{2}}{r} \right) + \boldsymbol{j}m \right).$$

The basic requirement is to find the phase factor ϕ which will make the amplitude nilpotent. So, we try the standard solution:

$$\phi = e^{-ar} r^{\gamma} \sum_{\nu=0} a_{\nu} r^{\nu}.$$

We then apply the operator we have defined to ϕ, and square the result to 0 to obtain:

$$4\left(E - \frac{A}{r} \right)^2 = -2\left(-a + \frac{\gamma}{r} + \frac{\nu}{r} + \cdots + \frac{1}{r} + i \frac{j + \frac{1}{2}}{r} \right)^2$$
$$- 2\left(-a + \frac{\gamma}{r} + \frac{\nu}{r} + \cdots + \frac{1}{r} - i \frac{j + \frac{1}{2}}{r} \right)^2 + 4m^2 .$$

Equating constant terms leads to

$$a = \sqrt{m^2 - E^2}.$$

Equating terms in $1/r^2$, following standard procedure, with $\nu = 0$, we obtain:

$$\left(\frac{A}{r} \right)^2 = -\left(\frac{\gamma + 1}{r} \right)^2 + \left(\frac{j + \frac{1}{2}}{r} \right)^2.$$

Assuming the power series terminates at n', following another standard procedure, and equating coefficients of $1/r$ for $\nu = n'$,

$$2EA = -2\sqrt{m^2 - E^2}\,(\gamma + 1 + n'),$$

the terms in $(j + 1/2)$ cancelling over the summation of the four multiplications, with two positive and two negative. Algebraic rearrangement of the last three equations then yields

$$\frac{E}{m} = \frac{1}{\sqrt{1 + \frac{A^2}{(\gamma+1+n')^2}}} = \frac{1}{\sqrt{1 + \frac{A^2}{\left(\sqrt{(j+1/2)^2 - A^2} + n'\right)^2}}},$$

With $A = Ze^2$, this is recognisable as the hyperfine or fine structure formula for a one-electron nuclear atom or ion (e.g. the hydrogen atom, where $Z = 1$).

6.6 The strong interaction

The strong interaction has never been explained analytically, though there is a general understanding, on empirical grounds, that it involves a linear potential. We have represented it as a nonlocal interaction between the parts of the wavefunction or 'quarks' incorporating p_x, p_y, and p_z. This interaction is independent of the physical separation of the components (and they must be spatially or temporally separated to have different local and nonlocal manifestations). The local effect of this is a transfer of $\mathbf{p}$ between the three brackets of each wavefunction. As it is local, with $\mathbf{p}$ changing within each bracket, it is not instantaneous, unlike the actual combinations and superpositions, and so there will be a nonzero local force or rate of change of momentum. However, because it does not depend on separation, this force will be constant. The local manifestation of a constant force is a potential that varies linearly with distance. In principle, it will be the same whether we are referring to a quark–antiquark combination with a temporal cycle of the components p_x, p_y, and p_z, and a separation of quark and antiquark, or to a combination of three quarks with a separation of each from the centre of the system. Using this argument based on fundamental principles, we create a differential operator incorporating Coulomb and linear potentials from a source with spherical symmetry, which is, in physical terms, either the centre of a 3-quark system or one component of a quark–antiquark pairing:

$$\left(\pm \boldsymbol{k}\left(E+\frac{A}{r}+Br\right) \mp \boldsymbol{i}\left(\frac{\partial}{\partial r}+\frac{1}{r} \pm i\frac{j+\frac{1}{2}}{r}\right)+\boldsymbol{ij}m\right).$$

The question is: can it be solved analytically using nilpotent methods? Again, we need to identify the phase factor to which the operator applies to yield nilpotent solutions. The pure Coulomb interaction gives us the template for solving more complicated systems, so, by analogy with this, we might propose that the phase factor is of the form:

$$\phi = \exp\left(-ar-br^2\right) r^{\gamma} \sum_{\nu=0} a_\nu r^\nu.$$

Applying the operator to this, and then the nilpotent condition, we obtain:

$$E^2+2AB+\frac{A^2}{r^2}+B^2r^2+\frac{2AE}{r}+2BEr$$
$$=m^2-\left(a^2+\frac{(\gamma+\nu+\,\cdots\,+1)^2}{r^2}-\frac{\left(j+\frac{1}{2}\right)^2}{r^2}+4b^2r^2+4abr\right.$$
$$\left.-4b\left(\gamma+\nu+\,\cdots\,+1\right)-\frac{2a}{r}\left(\gamma+\nu+\,\cdots\,+1\right)\right)$$

with the positive and negative $i(j+\frac{1}{2})$ terms cancelling out over the four solutions, as previously. Then, assuming a termination in the power series (as with the Coulomb solution), we can equate:

coefficients of r^2 to give	$B^2=-4b^2$
coefficients of r to give	$2BE=-4ab$
coefficients of $1/r$ to give	$2AE=2a\left(\gamma+\nu+1\right)$

These equations immediately lead to:

$$b=\pm\frac{iB}{2}$$
$$a=\mp iE$$
$$\gamma+\nu+1=\mp iA.$$

The ground state case (where $\nu=0$) then requires a phase factor of the form:

$$\phi=\exp\left(\pm iEr \mp iBr^2/2\right) r^{\mp iA-1}.$$

The imaginary exponential terms in ϕ can be seen as representing asymptotic freedom, the exp $(\mp iEr)$ being typical for a free fermion. The complex r^γ term can be structured as a component phase, $\chi(r) = \exp(\pm iA\ln(r))$, which varies less rapidly with r than the rest of ϕ. We can therefore write ϕ as

$$\phi = \frac{\exp\left(kr + \chi(r)\right)}{r},$$

where

$$k = \pm iE \mp iBr/2.$$

The first term in k dominates at high energies, where r is small, approximating to a free fermion solution, which can be interpreted as asymptotic freedom, while the second term, with its confining potential Br, dominates at low energies, when r is large, and this can be interpreted as infrared slavery. These are the established characteristics of the strong interaction and it seems that we have, for the first time, an explanation, derived on an analytic basis, for a force with these characteristics. The Coulomb term, which is required to maintain spherical symmetry, is the component which defines the strong interaction phase, $\chi(r)$, and this can be related to the directional status of $\mathbf{p}$ in the state vector. Once again, a nonlocal symmetry related to the nilpotent structure, determines the known characteristics of a local interaction.

6.7 The weak interaction

There does not seem to be any coherent idea of how the weak interaction should be expressed in terms of a distance-related potential, although its symmetry is known to be that of the $SU(2)$ group. We, however, have presented evidence for a dipole component, which, in more complicated situations, might be expected to become multipole. This time we will approach the problem more generally to show that the nilpotent solutions are exclusive. We will begin with a nilpotent operator with a form such as

$$\left(\boldsymbol{k}\left(E - \frac{A}{r} - Cr^n\right) + \boldsymbol{i}\left(\frac{\partial}{\partial r} + \frac{1}{r} \pm i\frac{j+\frac{1}{2}}{r}\right) + i\boldsymbol{j}m\right).$$

where n is an integer greater than 1 or less than -1 (so including the dipole–monopole and dipole–dipole cases of $n = -2$ and $n = -3$). As usual, we look for a phase factor which will make the amplitude nilpotent. Again, we turn to the Coulomb solution, with the additional information that polynomial potential terms which are multiples of r^n require the incorporation into the exponential of terms which are multiples of r^{n+1}. So, extending our work on the Coulomb solution, we may suppose that the phase factor is of the form:

$$\phi = \exp\left(-ar - br^{n+1}\right) r^{\gamma} \sum_{\nu=0} a_\nu r^\nu$$

Applying the operator and squaring to zero, with a termination in the series, we obtain

$$4\left(E - \frac{A}{r} - Cr^n\right)^2$$
$$= -2\left(-a - (n+1)\,br^n + \frac{\gamma}{r} + \frac{\nu}{r} + \frac{1}{r} + i\frac{j+\frac{1}{2}}{r}\right)^2$$
$$-2\left(-a - (n+1)\,br^n + \frac{\gamma}{r} + \frac{\nu}{r} + \frac{1}{r} - i\frac{j+\frac{1}{2}}{r}\right)^2 + 4m^2.$$

Equating constant terms, we find

$$a = \sqrt{m^2 - E^2}$$

Equating terms in r^{2n}, with $\nu = 0$:

$$C^2 = -(n+1)^2 b^2$$
$$b = \pm \frac{iC}{(n+1)}.$$

Equating coefficients of r^{n-1}, where $\nu = 0$:

$$AC = -(n+1)b(1+\gamma),$$
$$(1+\gamma) = \pm iA.$$

Equating coefficients of $1/r^2$ and coefficients of $1/r$, for a power series terminating in $\nu = n'$, we obtain

$$A^2 = -(1+\gamma+n')^2 + \left(j + \frac{1}{2}\right)^2$$

and

$$-EA = a(1+\gamma+n').$$

Combining the last three equations produces:

$$\left(\frac{m^2 - E^2}{E^2}\right)(1 + \gamma + n')^2 = -(1 + \gamma + n')^2 + \left(j + \frac{1}{2}\right)^2$$

$$E = -\frac{m}{j + \frac{1}{2}}(\pm iA + n') .$$

This equation has the form of a harmonic oscillator with evenly spaced energy levels deriving from integral values of n. Though it does not immediately suggest the value for the term iA, we can make the additional assumption, based on our interpretation of *zitterbewegung*, that A, the phase term required for spherical symmetry, has some connection with the random directionality of the fermion spin. We, therefore, assign to it a half-unit value $(\pm\frac{1}{2}i)$, or $(\pm\frac{1}{2}i\hbar c)$, using explicit values for the constants, and obtain the complete formula for the fermionic simple harmonic oscillator:

$$E = -\frac{m}{j + \frac{1}{2}}\left(\frac{1}{2} + n'\right).$$

The dimensions of A are those of charge (q) squared or interaction energy × range, and an A numerically equal to $\pm\frac{1}{2}\hbar c$ would be exactly that which is required by the uncertainty principle, allowing the value of the range of an interaction mediated by the Z boson to be calculated as $\hbar/2M_Zc = 2.166 \times 10^{-18}$ m, as observed. The $\frac{1}{2}\hbar c$ term is also significant in the expressions for zero-point energy and, as we have indicated, *zitterbewegung*, which connect with both spin and the uncertainty principle. Interpreting the *zitterbewegung* as a dipolar switching between fermionic and vacuum antifermionic states, we can describe this in terms of a weak dipole moment $(\hbar c/2)^{3/2}/M_Zc^2$, of magnitude $8.965 \times 10^{-18} e$ m (1.44×10^{-36} Cm). Because of the specific appearance of the $\frac{1}{2}\hbar c$ term for spin (s) in $\mu = gqs/2\,m$, an identical expression can additionally be used to define a weak magnetic moment of order $4.64 \times 10^{-5} \times$ the magnetic moment of the electron. The existence of such a dipole moment would make the spin $\frac{1}{2}$ term an expression of the dipolarity of the weak vacuum, and a physical representation of the weak interaction as a link between the fermion and vacuum, or between real space and vacuum space. The possible appearance of an imaginary factor i in A is interesting in relation to the requirement of a complex potential or vacuum for *CP* violation in the pure weak interaction.

The value of A we have assumed seems to be the one required to link a number of different physical manifestations of the weak interaction to fermionic structure. But, in any case, the solution we have derived indicates that an additional potential of the form Cr^n, where n is an integer greater than 1 or less than -1, has the effect of creating a harmonic oscillator solution for the nilpotent operator, irrespective of the value of n. In fact, we can show that any polynomial sum of potentials of this form will produce the same result, and consequently virtually any function of r other than the pure Coulomb of inverse linear, and the combination of inverse and direct linear. Potentials of this kind will emerge from any system in which there is complexity, aggregation, or a multiplicity of sources, even if the individual sources have Coulomb or linear potentials. In the case of dipolar weak sources, the minimum additional term will be of the form Cr^{-3}, and so will provide the correct characteristics for the weak interaction from the kind of potential that weak sources must necessarily produce. In addition, because this solution is exclusive for distance related potentials of the form Cr^n, except where $r = 1$ or -1, we have also, in effect, shown that a fermion interaction specified in relation to a spherically symmetric point source has only three physical manifestations, and that these are the ones associated with the electric (or other pure Coulomb), strong and weak interactions.

There is just one further aspect of the weak interaction that we need to resolve, certainly when it extends beyond the fermion-vacuum interaction to one between different fermionic states. The Coulomb interaction, as we have seen, manifestly obeys a $U(1)$ symmetry, and the $SU(3)$ symmetry for the strong interaction can be seen from the nilpotent structures we have identified for baryons. Can we establish that the symmetry group of the weak interaction as we have identified it is $SU(2)$? The dual state $\pm i\mathbf{k}E$ which produces the dipolar switching is, of course, manifestly an $SU(2)$ symmetry, and this can be related to the fact that the spherical symmetry of the point source proceeds from its independence from the handedness of the rotation, which, in terms of Noether's theorem, becomes the conservation of the handedness of angular momentum.

There are, however, *two different* $SU(2)$ symmetries involved, though, as we have previously hinted, they are related. The $SU(2)$ of spin describes two helicity states, left- and right-handed. However, *another* $SU(2)$ symmetry, weak isospin, describes a fermionic weak interaction as being independent of whether or not an electric charge is present and generating its own contribution to mass. The relation to the sign of $i\mathbf{k}E$ is based on the fact that the weakly interacting part with the positive sign of energy is purely left-handed, and the right-handed component has the 'wrong' sign of $\mathbf{p}$ relative to $i\mathbf{k}E$. Weak isospin, in effect, tells us that the weak interaction, though often occurring simultaneously and in combination with the electric interaction, has an independent origin. In this symmetry, the $SU(2)$ is a switching between different *ratios* of left- and right-handedness, and so of mass, determined by the presence or absence of the electric charge in one of the two states. So a superposition of, say, $\alpha_1(i\boldsymbol{k}E_1+\boldsymbol{i}\mathbf{p}_1+\boldsymbol{j}m_1)+\alpha_2(i\boldsymbol{k}E_1-\boldsymbol{i}\mathbf{p}_1+\boldsymbol{j}m_1)$ might become $\beta_1(i\boldsymbol{k}E_2+\boldsymbol{i}\mathbf{p}_2+\boldsymbol{j}m_2)+\beta_2(i\boldsymbol{k}E_2-\boldsymbol{i}\mathbf{p}_2+\boldsymbol{j}m_2)$, with both spin 1 and spin 0 vertices, the additional spin 0 vertex (the one which changes the mass) being equivalent to a fermion $(\pm i\boldsymbol{k}E\pm\boldsymbol{i}\mathbf{p}+\boldsymbol{j}m)$ acting on an 'electric' vacuum of the form $-\boldsymbol{j}(\pm i\boldsymbol{k}E\pm\boldsymbol{i}\mathbf{p}+\boldsymbol{j}m)\boldsymbol{j}$. In effect, a combination of the electric vacuum operator ($\boldsymbol{j}$) and the weak one ($\boldsymbol{k}$) produces a partial transition in the sign of $\mathbf{p}$, giving a basic indication of the way in which these forces are connected. (Gluons produce a direct transition in the strong interaction, and could be represented by terms like $(i\boldsymbol{k}E+\boldsymbol{i}ip_x)\boldsymbol{i}(i\boldsymbol{k}E+\boldsymbol{i}ip_x)\boldsymbol{i}$ reflecting the equivalence of the action on a nilpotent operator of a 'strong vacuum'.) It is significant that, while all the forces contribute to the production of the scalar mass term, the electric force is the only which contributes to nothing else.

The coupling of a massless fermion, say $(i\boldsymbol{k}E_1+\boldsymbol{i}\mathbf{p}_1)$, to a Higgs boson, say $(i\boldsymbol{k}E+\boldsymbol{i}\mathbf{p}+\boldsymbol{j}m)(-i\boldsymbol{k}E-\boldsymbol{i}\mathbf{p}+\boldsymbol{j}m)$, to produce a massive fermion, say $(i\boldsymbol{k}E_2+\boldsymbol{i}\mathbf{p}_2+\boldsymbol{j}m_2)$, can be imagined to be occurring at a vertex between the created fermion $(i\boldsymbol{k}E_2+\boldsymbol{i}\mathbf{p}_2+\boldsymbol{j}m_2)$ and the antistate $(i\boldsymbol{k}E_1-\boldsymbol{i}\mathbf{p}_1)$, the annihilated massless fermion, with subsequent equalisation of energy and momentum states. If we imagine a vertex involving a fermion superposing $(i\boldsymbol{k}E+\boldsymbol{i}\mathbf{p}+\boldsymbol{j}m)$ and $(i\boldsymbol{k}E-\boldsymbol{i}\mathbf{p}+\boldsymbol{j}m)$ with an antifermion superposing $(-i\boldsymbol{k}E+\boldsymbol{i}\mathbf{p}+\boldsymbol{j}m)$ and $(-i\boldsymbol{k}E-\boldsymbol{i}\mathbf{p}+\boldsymbol{j}m)$, then there will be a minimum of two spin 1 combinations and two spin 0 combinations, meaning that the vertex will be massive (with Higgs coupling) and carry a non-weak (i.e. electric) charge. So, a process such as a weak isospin transition, which, to use a very basic model, converts something like $(i\boldsymbol{k}E_1+\boldsymbol{i}\mathbf{p}_1+\boldsymbol{j}m_1)$ (representing isospin up) to something like $\alpha_1\ (i\boldsymbol{k}E_2+\boldsymbol{i}\mathbf{p}_2+\boldsymbol{j}m_2)+\alpha_2(i\boldsymbol{k}E_2-\boldsymbol{i}\mathbf{p}_2+\boldsymbol{j}m_2)$ (representing isospin down), requires an additional Higgs boson vertex (spin 0) to accommodate the right-handed part of the isospin down state, when the left-handed part interacts weakly. This is, of course, what we mean when we say that the W and Z bosons have mass. The mass balance is done through separate vertices involving the Higgs boson.

We have now seen that the nilpotent structure, with its pseudoscalar, vector and scalar components, *already incorporates the fundamental interactions.* A nilpotent fermion defined by this mathematical formalism is *necessarily* acting according to some or all of these interactions. They arise solely from its internal structure. Coulomb terms are simply the result of the spherical symmetry of point sources. Since the Coulomb interaction is purely an expression of the magnitude of a scalar phase, all the terms in the nilpotent contribute, but only one, the passive (scalar) mass term, contributes to nothing else, and that one term in the nilpotent operator has no structure other than magnitude, which ensures that it must be possible to have an interaction with no symmetry other than $U(1)$. An interaction with this precise property may therefore be defined, and it is the one we define as the *electric* interaction. At the same time, the strong interaction, with its characteristic linear potential, can be represented as we have seen, by the vector properties of the $\mathbf{p}$ term. It may be significant that the linear potential of the strong interaction is the only one that is optional to the fermionic state, the nilpotency not being dependent directly on the vector nature of $\mathbf{p}$.

However, yet another interaction is required by the *spinor* structure of the nilpotent operator and the associated phenomenon of *zitterbewegung*. While the co-existence of two spin states is, in some sense, real and is accounted for by the presence of mass, the co-existence of two energy states is only meaningful in the context of the simultaneous existence of fermion and vacuum. While the transitions between the two energy states may be virtual in this sense, the *zitterbewegung* would seem to require the production of an intermediate bosonic state at a vertex where one fermionic state is annihilated and another is created to replace it. This behaviour is, of course, characteristic of the weak interaction, and, in this sense, we can say that the weak interaction, like the electric and strong interactions, is built into the structure of the nilpotent operator, and its nature is determined by that of the pseudoscalar iE operator, whose sign uniquely determines the helicity of a weakly interacting particle or more specifically, its weakly interacting component.

Ultimately, the broken symmetry between the interactions, which is manifested in these structures, emerges when the 8 base units for time, space, mass and charge are compactified into 5 composite generators of the group of order 64 which they construct. Both time, space and mass, and simultaneously charge, are modified in the process. The modification of charge shows the nonlocal or vacuum side of the compactification process, while the compactification to energy, momentum and rest mass show the local. The algebraic characteristics acquired are manifested nonlocally through the vacua associated with the energy, momentum, and rest mass components of the nilpotent wavefunction. The different algebraic characteristics of the three components then ultimately determine the nature of the local interactions which result from the local symmetries. We have shown that it is possible to go all the way from the nonlocal vacua to the recognisable physical effects which are characteristic of the different forces.

Chapter 7

Nilpotent Quantum Field Theory

7.1 A perturbation calculation

We are now ready to extend the discussion to a yet higher level of complexity to show that fundamental considerations still apply there, and lead us to results not so far achieved by any other method. Some of the calculations will be more involved than the ones we have done previously, but they are necessary to show how far the fundamental ideas can be influential, even at a level of relative complexity, and can solve problems for which there are no current answers. These will be followed by a less technical consideration of some more general aspects of the nilpotent quantum theory.

A significant amount of quantum field theory is already present in nilpotent quantum mechanics, where the nilpotent operator already provides interaction over the entire quantum field. Nilpotent 'wavefunctions' are the result of creation and annihilation operators acting on the vacuum state and come already second quantized. The mathematical proof of this is given in *Zero to Infinity*, in addition to results in QED, including an electron scattering calculation and a derivation of renormalization for interacting particles.[10] Renormalization, in the sense of rescaling, as the interacting electric charges are screened by the vacuum field, should still apply as in conventional QED. What should no longer apply, if the nilpotent formalism is a more fundamentally symmetric — and in this sense more 'natural' — version of quantum mechanics, is the need for the infinite self-energy term that has caused such problems in the past. If the nilpotent formalism derives from the symmetries which are the most fundamental in nature, then something of this kind should only arise as an artefact of alternative mathematical structures in which these symmetries are not fully preserved.

In the last chapter, we saw that an exact supersymmetry appears as a consequence of the nilpotent formalism and its representation of vacuum. In this case, we should expect a free fermion in vacuum to produce its own loop cancellations and its energy to acquire a finite value without renormalization. Free fermion plus boson loops should cancel at all levels of calculation, and there should be no hierarchy problem. We can examine this possibility by performing a basic perturbation calculation for first order coupling in QED, and seeing if it leads to zero in the case of a free fermion. Let us suppose we have a fermion acted on by the electromagnetic potentials ϕ, $\mathbf{A}$. Then, using only the lead terms of the spinors for simplicity, we have the standard equation

$$\left(-\left(\boldsymbol{k}\frac{\partial}{\partial t} - i\boldsymbol{k}e\phi\right) - (\boldsymbol{i}\nabla - \boldsymbol{i}ie\mathbf{A}) + \boldsymbol{j}m\right)\psi = 0,$$

which can be rearranged as

$$\left(-\boldsymbol{k}\frac{\partial}{\partial t} - i\boldsymbol{i}\nabla + \boldsymbol{j}m\right)\psi = -e(i\boldsymbol{k}\phi + i\boldsymbol{i}\mathbf{A})\psi$$

We now apply a perturbation expansion to ψ, so that

$$\psi = \psi_0 + \psi_1 + \psi_2 + \cdots,$$

with

$$\psi_0 = (i\boldsymbol{k}E + \boldsymbol{i}\mathbf{p} + \boldsymbol{j}m)e^{-i(Et - \mathbf{p}\cdot\mathbf{r})}$$

as the solution of the unperturbed equation:

$$\left(-\boldsymbol{k}\frac{\partial}{\partial t} - i\boldsymbol{i}\nabla + \boldsymbol{j}m\right)\psi = 0,$$

which represents zeroth-order coupling, or a free fermion of momentum $\mathbf{p}$.

Using the perturbation expansion, we can write

$$\left(-\boldsymbol{k}\frac{\partial}{\partial t} - i\boldsymbol{i}\nabla + \boldsymbol{j}m\right)(\psi_0 + \psi_1 + \psi_2 + \cdots)$$
$$= -e(i\boldsymbol{k}\phi + i\boldsymbol{i}\mathbf{A})(\psi_0 + \psi_1 + \psi_2 + \cdots),$$

from which we can extract the first-order coupling from the first iteration of the perturbation expansion, as

$$\left(-\boldsymbol{k}\frac{\partial}{\partial t} - i\boldsymbol{i}\nabla + \boldsymbol{j}m\right)\psi_1 = -e(i\boldsymbol{k}\phi + i\boldsymbol{i}\mathbf{A})\psi_0.$$

If, using a standard technique, we expand $(\boldsymbol{ik}\phi + \boldsymbol{ii}\mathbf{A})$ as a Fourier series, and sum over momentum $\mathbf{k}$, we obtain

$$(\boldsymbol{ik}\phi + \boldsymbol{ii}\mathbf{A}) = \sum (\boldsymbol{ik}\phi(\mathbf{k}) + \boldsymbol{ii}\mathbf{A}(\mathbf{k}))e^{i\mathbf{k}\cdot\mathbf{r}},$$

so that

$$\left(-\boldsymbol{k}\frac{\partial}{\partial t} - \boldsymbol{ii}\nabla + \boldsymbol{j}m\right)\psi_1$$
$$= -e\sum (\boldsymbol{ik}\phi(\mathbf{k}) + \boldsymbol{ii}\mathbf{A}(\mathbf{k}))e^{i\mathbf{k}\cdot\mathbf{r}}\psi_0$$
$$= -e\sum (\boldsymbol{ik}\phi(\mathbf{k}) + \boldsymbol{ii}\mathbf{A}(\mathbf{k}))e^{i\mathbf{k}\cdot\mathbf{r}}(\boldsymbol{ik}E + \boldsymbol{i}\mathbf{p} + \boldsymbol{j}m)e^{-i(Et-\mathbf{p}\cdot\mathbf{r})}$$
$$= -e\sum (\boldsymbol{ik}\phi(\mathbf{k}) + \boldsymbol{ii}\mathbf{A}(\mathbf{k}))(\boldsymbol{ik}E + \boldsymbol{i}\mathbf{p} + \boldsymbol{j}m)e^{-i(Et-(\mathbf{p}+\mathbf{k})\cdot\mathbf{r})}.$$

If we now expand ψ_1 as

$$\psi_1 = \sum \nu_1(E, \mathbf{p}+\mathbf{k})e^{-i(Et-(\mathbf{p}+\mathbf{k})\cdot\mathbf{r})}$$

then

$$\sum \left(-\boldsymbol{k}\frac{\partial}{\partial t} - \boldsymbol{ii}\nabla + \boldsymbol{j}m\right)\nu_1(E, \mathbf{p}+\mathbf{k})e^{-i(Et-(\mathbf{p}+\mathbf{k})\cdot\mathbf{r})}$$
$$= -e\sum (\boldsymbol{ik}\phi(\mathbf{k}) + \boldsymbol{ii}\mathbf{A}(\mathbf{k}))(\boldsymbol{ik}E + \boldsymbol{i}\mathbf{p} + \boldsymbol{j}m)e^{-i(Et-(\mathbf{p}+\mathbf{k})\cdot\mathbf{r})}$$

and

$$\sum (\boldsymbol{ik}E + \boldsymbol{i}(\mathbf{p}+\mathbf{k}) + \boldsymbol{j}m)\nu_1(E, \mathbf{p}+\mathbf{k})e^{-i(Et-(\mathbf{p}+\mathbf{k})\cdot\mathbf{r})}$$
$$= -e\sum (\boldsymbol{ik}\phi(\mathbf{k}) + \boldsymbol{ii}\mathbf{A}(\mathbf{k}))(\boldsymbol{ik}E + \boldsymbol{i}\mathbf{p} + \boldsymbol{j}m)e^{-i(Et-(\mathbf{p}+\mathbf{k})\cdot\mathbf{r})}$$

and, equating individual terms,

$$(\boldsymbol{ik}E + \boldsymbol{i}(\mathbf{p}+\mathbf{k}) + \boldsymbol{j}m)v_1(E, \mathbf{p}+\mathbf{k}) = -e(\boldsymbol{ik}\phi(\mathbf{k}) + \boldsymbol{ii}\mathbf{A}(\mathbf{k}))(\boldsymbol{ik}E + \boldsymbol{i}\mathbf{p} + \boldsymbol{j}m).$$

We can write this in the form

$$v_1(E, \mathbf{p}+\mathbf{k}) = -e[\boldsymbol{ik}E + \boldsymbol{i}(\mathbf{p}+\mathbf{k}) + \boldsymbol{i}\boldsymbol{j}m]^{-1}(\boldsymbol{ik}\phi(\mathbf{k}) + \boldsymbol{ii}\mathbf{A}\ (\mathbf{k}))(\boldsymbol{ik}E + \boldsymbol{i}\mathbf{p} + \boldsymbol{j}m)$$

which means that

$$\psi_1 = -e\sum [\boldsymbol{ik}E + \boldsymbol{i}(\mathbf{p}+\mathbf{k}) + \boldsymbol{j}m]^{-1}(\boldsymbol{ik}\phi(\mathbf{k}) + \boldsymbol{ii}\mathbf{A}(\mathbf{k}))$$
$$\times(\boldsymbol{ik}E + \boldsymbol{i}\mathbf{p} + \boldsymbol{j}m)e^{-i(Et-(\mathbf{p}+\mathbf{k})\cdot\mathbf{r})}$$

This is the wavefunction for first-order coupling, with a fermion absorbing or emitting a photon of momentum $\mathbf{k}$.

But, if we observe the process in the rest frame of the fermion and eliminate any *external* source of potential, then $\mathbf{k} = 0$, and the only possible potential $(\boldsymbol{ik}\phi + \boldsymbol{i}i\mathbf{A})$ that could apply is the internal, self-interacting one not dependent on $\mathbf{k}$, which, in the rest frame, will reduce to the static value, $\boldsymbol{ik}\phi$, with ϕ as a self-potential. In this case, ψ_1 becomes

$$\psi_1 = -e[\boldsymbol{ik}E + \boldsymbol{i}\mathbf{p} + \boldsymbol{j}m]^{-1}(\boldsymbol{ik}\phi)(\boldsymbol{ik}E + \boldsymbol{i}\mathbf{p} + \boldsymbol{j}m)e^{-i(Et-\mathbf{p}\cdot\mathbf{r})},$$

as the summation is no longer strictly required for a single order of the pure self-interaction. Since we can also write this as

$$\psi_1 = -e(-\boldsymbol{ik}E + \boldsymbol{i}\mathbf{p} + \boldsymbol{j}m)(-\boldsymbol{ik}E + \boldsymbol{i}\mathbf{p} + \boldsymbol{j}m)(\boldsymbol{ik}\phi)e^{-i(Et-\mathbf{p}\cdot\mathbf{r})},$$

we see that $\psi_1 = 0$, for any fixed value of ψ. Clearly, this will also apply to higher orders of self-interaction. In other words, we have a first indication that a *non-interacting* nilpotent fermion requires no renormalization as a result of its self-energy.

The process could also be adapted for interacting particles subject to external potentials. Here we can imagine redefining the E and $\mathbf{p}$ operators to incorporate external potentials to make them 'internal', while simultaneously changing the structure of the phase factor to accommodate this. The change of phase factor would, of course, require a corresponding change in the amplitude, which could be taken as redetermining the value of the coupling constant, e, as required. Ultimately, however, it is the structure of $(\boldsymbol{ik}E + \boldsymbol{i}\mathbf{p} + \boldsymbol{j}m)$ as a *nilpotent* which seemingly eliminates the infinite self-interaction terms in the perturbation expansion at the same time as showing that they are merely an expression of the nature of the nilpotent vacuum as a reflection of the exactly supersymmetric nature of the original particle state.

7.2 Cancellation of loops

If the previous argument is correct, then we should also be able to achieve the same result using the supersymmetric properties of the nilpotent operator to cancel fermion and boson loops directly. This is precisely what we would expect from a nilpotent system where the total energy is zero, and one way of realising this would be to combine negative energy fermions with positive energy bosons. In the nilpotent formulation, as we have seen, every fermionic state has an intrinsic supersymmetric spin 1 bosonic vacuum partner with the same energy, momentum and mass. If we represent a spin $\frac{1}{2}$ fermion by, say, $(\pm i\boldsymbol{k}E \pm \boldsymbol{i}\mathbf{p} + \boldsymbol{j}m)$, and a spin $-\frac{1}{2}$ fermion by $(\pm i\boldsymbol{k}E \mp \boldsymbol{i}\mathbf{p} + \boldsymbol{j}m)$, then each of these is unchanged by postmultiplication any number of times by the vacuum operator $\boldsymbol{k}(\pm i\boldsymbol{k}E \pm \boldsymbol{i}\mathbf{p} + \boldsymbol{j}m)$ or $\boldsymbol{k}(\pm i\boldsymbol{k}E \mp \boldsymbol{i}\mathbf{p} + \boldsymbol{j}m)$.

However,

$$(\pm i\boldsymbol{k}E \pm \boldsymbol{i}\mathbf{p} + \boldsymbol{j}m)\boldsymbol{k}(\pm i\boldsymbol{k}E \pm \boldsymbol{i}\mathbf{p} + \boldsymbol{j}m) \times \boldsymbol{k}(\pm i\boldsymbol{k}E \pm \boldsymbol{i}\mathbf{p} + \boldsymbol{j}m)\boldsymbol{k}(\pm i\boldsymbol{k}E \pm \boldsymbol{i}\mathbf{p} + \boldsymbol{j}m)\cdots$$

and

$$(\pm i\boldsymbol{k}E \mp \boldsymbol{i}\mathbf{p} + \boldsymbol{j}m)\boldsymbol{k}(\pm i\boldsymbol{k}E \mp \boldsymbol{i}\mathbf{p} + \boldsymbol{j}m) \times \boldsymbol{k}(\pm i\boldsymbol{k}E \mp \boldsymbol{i}\mathbf{p} + \boldsymbol{j}m)\boldsymbol{k}(\pm i\boldsymbol{k}E \mp \boldsymbol{i}\mathbf{p} + \boldsymbol{j}m)\cdots$$

are indistinguishable from

$$(\pm i\boldsymbol{k}E \pm \boldsymbol{i}\mathbf{p} + \boldsymbol{j}m)(\mp i\boldsymbol{k}E \pm \boldsymbol{i}\mathbf{p} + \boldsymbol{j}m) \times (\pm i\boldsymbol{k}E \pm \boldsymbol{i}\mathbf{p} + \boldsymbol{j}m)(\mp i\boldsymbol{k}E \pm \boldsymbol{i}\mathbf{p} + \boldsymbol{j}m)\cdots$$

and

$$(\pm i\boldsymbol{k}E \mp \boldsymbol{i}\mathbf{p} + \boldsymbol{j}m)(\mp i\boldsymbol{k}E \mp \boldsymbol{i}\mathbf{p} + \boldsymbol{j}m) \times (\pm i\boldsymbol{k}E \mp \boldsymbol{i}\mathbf{p} + \boldsymbol{j}m)(\mp i\boldsymbol{k}E \mp \boldsymbol{i}\mathbf{p} + \boldsymbol{j}m)\cdots$$

which alternate spin $\frac{1}{2}$ and spin $-\frac{1}{2}$ fermions with spin 1 and spin -1 bosons. In effect the fermion generates its own vacuum boson partner, with the same E, $\mathbf{p}$ and m. Since the nilpotent structure is founded on zero totality, with the vacuum and fermion being in both zero superposition and zero combination, we may assume that this is an indication that the total energy made by positive boson and negative fermion loops is zero.

The calculation is surprisingly easy if we use results obtained from conventional QED. In fact, we can reduce it to simple arithmetic! Using a result from a standard textbook,[11] we find that the vacuum energy for a particle of mass m and spin j is given by:

$$\frac{1}{2}(-1)^{2j}(2j+1)\int d^3k\sqrt{k^2+m_j^2} = \frac{1}{2}(-1)^{2j}(2j+1)\int d^3k\sqrt{k^2}\left(1+\frac{1}{2}\frac{m_j^2}{k^2}-\frac{1}{8}\left(\frac{m_j^2}{k^2}\right)^2+\cdots\right)$$

Here we see quartic, quadratic and logarithmic divergences. To remove these, we need to ensure that

$$\sum_j (-1)^{2j}(2j+1) = 0$$

$$\sum_j (-1)^{2j}(2j+1)m_j^2 = 0$$

$$\sum_j (-1)^{2j}(2j+1)m_j^4 = 0$$

The first condition requires equal numbers of fermionic and bosonic degrees of freedom. If we have $j = \pm\frac{1}{2}$ for the fermionic loops and $j = \pm 1$ for the bosonic loops, then

$$(-1)^{2j}(2j+1) = -2 \quad \text{for } j = \frac{1}{2}$$

$$(-1)^{2j}(2j+1) = 3 \quad \text{for } j = 1$$

$$(-1)^{2j}(2j+1) = 0 \quad \text{for } j = -\frac{1}{2}$$

$$(-1)^{2j}(2j+1) = -1 \quad \text{for } j = -1$$

giving a total of

$$\sum_j (-1)^{2j}(2j+1) = -2 + 3 + 0 - 1 = 0$$

as required.

The other two conditions additionally require the fermions and bosons to have equal masses, which is true if the supersymmetry is intrinsic. Since all three conditions are fulfilled in the nilpotent formalism, it would appear that the intrinsic supersymmetry automatically removes the ultraviolet divergence.

The same hierarchy problem of divergence at each level of calculation also applies to bosons, most famously in the case of the spin 0 Higgs boson, but the same reasoning should also apply here. For a spin 0 boson, we have a fundamental structure of either

$$(\pm i\boldsymbol{k}E \pm \mathbf{ip} + \boldsymbol{j}m)(\mp i\boldsymbol{k}E \mp \mathbf{ip} + \boldsymbol{j}m)$$

or

$$(\pm i\boldsymbol{k}E \mp \mathbf{ip} + \boldsymbol{j}m)(\mp i\boldsymbol{k}E \pm \mathbf{ip} + \boldsymbol{j}m)$$

with a combination of spin $\frac{1}{2}$ and spin $-\frac{1}{2}$ fermions/antifermions (to which we can again apply vacuum operators). (The application of vacuum operators to the two partners in the combination would leave alternate creations of fermion and boson as before.) Since

$$(-1)^{2j}(2j+1)^{2j} = 1 \quad \text{for } j = 0$$

we can find a combination of spin $\frac{1}{2}$ and spin 0, together with spin $-\frac{1}{2}$ and spin 0, which will lead to

$$\sum_j (-1)^{2j}(2j+1) = -2 + 1 + 0 + 1 = 0$$

again as required, and, with m common to fermions and bosons, it also fulfills the second and third conditions. It would appear from this argument that the divergence is again removed and, in particular, that there is no reason to expect a hierarchy problem for the Higgs boson.

An additional related problem is the matter/antimatter asymmetry between fermions and antifermions. Answers to this long-standing problem have been generally sought for in cosmology. It is assumed that almost equal amounts of fermions and antifermions were created in the Big Bang, with the fermions *slightly* in excess. Following this, the mysterious process of baryogenesis led to the annihilation of all the antifermions. But, there are very good reasons for seeing the asymmetry as generic, and, in fact, not an asymmetry at all. According to our foundational ideas, we have two vector spaces, characterised in the nilpotent representation by positive and negative energies. Between these two spaces, there are the same number of fermions and antifermions. Just as fermions, with E, can be seen as the characteristic particles defining real (observable) space, antifermions, with $-E$, can be seen as the characteristic particles defining vacuum space. In this description, there will not be a symmetry between the two particle types in either of the spaces. In the nilpotent structure, there are two energies, two directions of time, two directions of spin, two fermions and two antifermions. There are even two causalities: forward causality for the local state, the thing we observe; and *backward* or reverse causality for the unobservable nonlocal vacuum, which contains all the future causes of everything that will happen. The first corresponds with E and t, the second with $-E$ and $-t$. Every fundamental concept provides us with a totality zero. Only rest mass has a purely positive value, but this is not fundamental, being a concept whose main purpose is to separate the observational part of the structure from that which is not observed.

7.3 Propagators

An aspect of quantum field theory which benefits massively from being cast in the nilpotent, or, we could say, *fundamental*, formalism is the use of *propagators* (which are, in principle, Green's functions). Though this is a rather technical subject, it does show the power of the nilpotent concepts in a particularly direct way, and also shows the significance of the dual spaces in creating fermions as point singularities. Here, again, there is a divergence which is not fundamental, but which we can show to be a result of using a less symmetrical mathematical formalism. In addition, the concept of a boson propagator has not been fully worked out, as there are in fact three boson propagators corresponding to the three different types of boson, and not a generic one which doesn't quite correspond to any of them.

We have seen that a physical singularity (perhaps the only one that can exist) emerges from the combination of two dual vector spaces at the same time as a *zitterbewegung* is generated through the switching between them, which is equivalent to the switching between $+iE$ and $-iE$. Now, the conventional Feynman formalism for the particle propagator produces a 'pole' or singularity at exactly the division or 'switchover' between these two energy states, or between fermion and antifermion. It is a problem because it leads to a divergence at that point, which can only be dealt with by a mathematical subterfuge. In the nilpotent theory, however, the pole is no longer a 'naked' singularity, causing an infinite divergence, but one that is accommodated within the dual spaces on which the theory is founded. The nilpotent formalism incorporates the pole automatically without divergence because of its direct inclusion of vacuum states.

Conventional theory assumes that a fermion propagator takes the form

$$S_F(p) = \frac{1}{\not{p} - m} = \frac{\not{p} + m}{p^2 - m^2},$$

where $\not{p}$ represents $\gamma^\mu \partial_\mu$, or its eigenvalue, and that there is a singularity or 'pole' (p_0) where $p^2 - m^2 = 0$, the 'pole' being the origin of antifermion states. On either side of the pole there are positive energy states moving forwards in time, and negative energy states moving backwards in time, the terms $(\not{p} + m)$ and $(-\not{p} + m)$ being used to project out, respectively, the positive and negative energy states. The normal solution is to add an infinitesimal term $i\varepsilon$ to $p^2 - m^2$, so that $S_F(p)$ becomes

$$S_F(p) = \frac{1}{\not{p} - m} = \frac{1}{2p_0}\left(\frac{1}{p_0 - \sqrt{p^2 + m^2} + i\varepsilon} + \frac{1}{p_0 + \sqrt{p^2 + m^2} - i\varepsilon}\right)$$

and take a contour integral over the complex variable to give the solution

$$S_F(x-x') = \int d^3p \frac{1}{(2\pi)^3}\frac{m}{2E} \times \left[-i\theta(t-t')\sum_{r=1}^{2}\Psi(x)\bar{\Psi}(x') + i\theta(t'-t)\sum_{r=3}^{4}\Psi(x)\bar{\Psi}(x')\right]$$

with summations over the up and down spin states.

This mathematical subterfuge is unnecessary in the nilpotent formalism because the denominator of the propagator term is always a nonzero scalar. We write

$$S_F(p) = \frac{1}{(\pm i\boldsymbol{k}E \pm i\mathbf{p} + \boldsymbol{j}m)},$$

and choose our usual interpretation of the reciprocal of a nilpotent to give:

$$\frac{1}{(\pm i\boldsymbol{k}E \pm i\mathbf{p} + \boldsymbol{j}m)} = \frac{(\pm i\boldsymbol{k}E \mp i\mathbf{p} - \boldsymbol{j}m)}{(\pm i\boldsymbol{k}E \pm i\mathbf{p} + \boldsymbol{j}m)\,(\pm i\boldsymbol{k}E \mp i\mathbf{p} - \boldsymbol{j}m)} = \frac{(\pm i\boldsymbol{k}E \mp i\mathbf{p} - \boldsymbol{j}m)}{4\,(E^2+p^2+m^2)},$$

which is finite at all values. The integral is now simply

$$S_F(x-x') = \int d^3p \frac{1}{(2\pi)^3}\frac{m}{2E}\theta(t-t')\Psi(x)\bar{\Psi}(x'),$$

in which $\Psi(x)$ is the usual

$$\Psi(x) = (\pm i\boldsymbol{k}E \pm \boldsymbol{i}\mathbf{p} + \boldsymbol{j}m)\exp(ipx),$$

with the phase factor written as a 4-vector, and the adjoint term becomes

$$\bar{\Psi}(x') = (\mp i\boldsymbol{k}E \pm \boldsymbol{i}\mathbf{p} + \boldsymbol{j}m)(i\boldsymbol{k})\exp(-ipx').$$

Since the nilpotent formalism comes as a complete package with a single phase term, automatic second quantization, and the negative energy states matched with reverse time states, there is no averaging over spin states or separation of positive and negative energy states on opposite sides of a pole. The particle structure is itself the singularity. There is no division between the particle and antiparticle because the two come as a single unit incorporating real space and vacuum space on an equal footing.

The fermion propagator can also be used to define boson propagators. In conventional theory, we derive the boson propagator directly from the Klein–Gordon equation, while recognising that its mathematical form depends on the choice of gauge:

$$\Delta_F(x - x') = \frac{\not{p} + m}{p^2 - m^2}.$$

This is because the Klein–Gordon operator

$$\left(\gamma^0 \frac{\partial}{\partial t} + \gamma \cdot \nabla + im\right)\left(\gamma^0 \frac{\partial}{\partial t} + \gamma \cdot \nabla - im\right) = \left(\frac{\partial^2}{\partial t^2} - \nabla^2 + m^2\right)$$

is the only scalar product which can emerge from a linear differential operator defined as in the two bracketed terms on the left. The Klein–Gordon equation, however, is not specific to bosonic states or an identifier of them. It merely defines a universal zero condition which is true for all states, whether bosonic or fermionic. And, the propagator defined by conventional theory does not correspond to any known bosonic state. Instead, we have *three* boson propagators.

$$\text{Spin 1: } \Delta_F(x - x') = \frac{1}{(\pm i\boldsymbol{k}E \pm i\mathbf{p} + \boldsymbol{j}m)(\mp i\boldsymbol{k}E \pm i\mathbf{p} + \boldsymbol{j}m)},$$

$$\text{Spin 0: } \Delta_F(x - x') = \frac{1}{(\pm i\boldsymbol{k}E \pm i\mathbf{p} + \boldsymbol{j}m)(\mp i\boldsymbol{k}E \mp i\mathbf{p} + \boldsymbol{j}m)},$$

$$\text{Paired Fermion: } \Delta_F(x - x') = \frac{1}{(\pm i\boldsymbol{k}E \pm i\mathbf{p} + \boldsymbol{j}m)(\pm i\boldsymbol{k}E \mp i\mathbf{p} + \boldsymbol{j}m)}.$$

Where the spin 1 bosons are massless (as in QED), we will have expressions like:

$$\Delta_F(x - x') = \frac{1}{(\pm i\boldsymbol{k}E \pm i\mathbf{p})(\mp i\boldsymbol{k}E \pm i\mathbf{p})}.$$

Clearly, the relationship of the fermion and boson propagators is of the form

$$S_F(x - x') = (i\gamma^\mu \partial_\mu + m)\Delta_F(x - x'),$$

or, in our notation,

$$S_F(x - x') = (\pm i\boldsymbol{k}E \pm i\mathbf{p} + \boldsymbol{j}m)\Delta_F(x - x'),$$

which is exactly the same relationship as is defined between fermion and boson in the nilpotent formalism. Now, using

$$S_F(p) = \frac{1}{2p_0}\left(\frac{1}{p_0 - \sqrt{p^2+m^2} + i\varepsilon} + \frac{1}{p_0 + \sqrt{p^2+m^2} - i\varepsilon}\right),$$

which is the same as the conventional fermion propagator up to a factor $(\not{p} + m)$, we can perform a contour integral which is similar to that for the fermion to produce

$$i\Delta_F(x - x') = \int d^3p \frac{1}{(2\pi)^3} \frac{1}{2\omega} \theta(t - t')\phi(x)\phi^*(x').$$

Here, ω takes the place of E/m, while $\phi(x)$ and $\phi(x')$ are now scalar wavefunctions. However, in our notation, they will be scalar products of $(\pm i\boldsymbol{k}E \pm \mathbf{ip} + \boldsymbol{j}m)\exp(ipx)$ and $(\mp i\boldsymbol{k}E \pm \mathbf{ip} + \boldsymbol{j}m)\exp(ipx')$ and $\phi(x)\phi^*(x')$ reduces to a product of a scalar term, which can be removed by normalization, and exp $ip(x - x')$.

In off-mass-shell conditions, where $E^2 \neq p^2 + m^2$, poles in the propagator are a mathematical, rather than a physical, problem, and are removed by the use of $i\varepsilon$ and the contour integral, which is *ad hoc* but effective. But, in the specific case of massless bosons, such as the photon or gluon, conventional theory cannot prevent 'infrared' divergences appearing in the expression for the propagator when such bosons are emitted from an initial or final stage which is on the mass shell. Such divergences, however, do not occur where there is no naked pole, as in the nilpotent expression. The nilpotent definition of the boson propagator not only shows that one of the principal divergences in quantum electrodynamics is, as the procedure used to remove it would suggest, merely an artefact of the mathematical structure we have imposed, and not of a fundamentally physical nature, but also suggests that the formalism which removes it is a more exact representation of the fundamental physics. Ultimately, this is because it allows an exact representation of the vacuum simultaneously with the fermionic state, in line with the dual spaces needed to generate a fermion singularity.

7.4 A weak interaction calculation

To complete the more technical aspects of nilpotent quantum field theory, we can show how it would be used in a weak interaction calculation in the usual four-point Fermi interaction approximation. Though there is nothing here that can't be done by conventional methods, it is interesting to see how a different approach could possibly be more fertile in more complex problems. Conventionally, in describing a weak interaction, such as muon decay, we calculate the traces of the tensors using the trace theorem:

$$Tr[\gamma^\mu(1-\gamma^5)p_1\gamma^\nu(1-\gamma^5)p_3]\,Tr[\gamma_\mu(1-\gamma^5)p_2\gamma_\nu(1-\gamma^5)p_4]$$
$$= 256(p_1{\cdot}p_2)(p_3{\cdot}p_4)$$

This is because, for an invariant amplitude M, for muon decay,

$$|\mathcal{M}|^2 = \frac{G^2}{2}Tr[\gamma_\mu(1-\gamma^5)\bar{\gamma}_\mu(p_1)\gamma^\nu(1-\gamma^5)\mu(p_3)]$$
$$\times Tr[\bar{\nu}_e(1-\gamma^5)(p_2)e(1-\gamma^5)(p_4)]$$

and the spin-averaged probability, $|\mathcal{M}|^2$, is given by

$$\frac{1}{2}\sum_{\text{spins}} Tr[\gamma^\mu(1-\gamma^5)\bar{\nu}_\mu(p_1)\gamma^\nu(1-\gamma^5)\mu(p_3)]$$
$$\times Tr[\gamma_\mu(1-\gamma^5)\bar{\nu}_e(p_2)\gamma_\nu(1-\gamma^5)e(p_4)]$$
$$= 64G^2(p_1\cdot p_2)(p_3\cdot p_4)$$

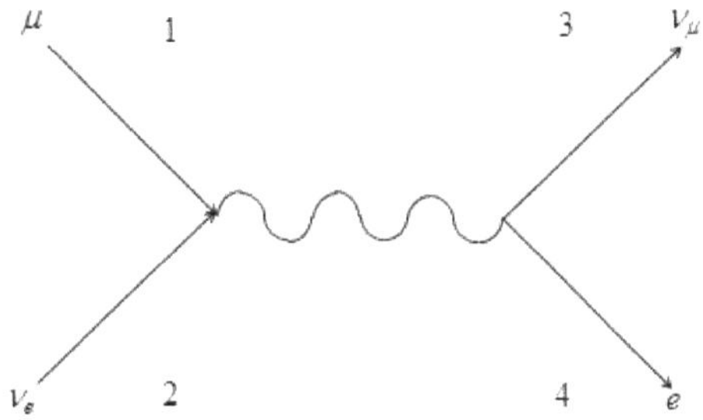

Using nilpotents we can take a different approach, by directly investigating the 'bosonic' states at the two vertices. First we take:

$$\gamma^\mu = \gamma^0 + \gamma^1 + \gamma^2 + \gamma^3 = i\boldsymbol{k} + \mathbf{i}\boldsymbol{i} + \mathbf{j}\boldsymbol{i} + \mathbf{k}\boldsymbol{i} = i\boldsymbol{k} + \mathbf{1}\boldsymbol{i},$$

combining the vectors **i**, **j**, **k**, for convenience, into the single vector symbol **1**. Then:

$$(1-\gamma^5) = (1 - i\boldsymbol{j})$$
$$\gamma^\mu(1-\gamma^5) = (i\boldsymbol{k} + \mathbf{1}\boldsymbol{i})(1 - i\boldsymbol{j}) = i\boldsymbol{k} \;\; - \boldsymbol{i} + \mathbf{1}\boldsymbol{i} - i\boldsymbol{k}\mathbf{1}$$

Here, we assume:

$$p_1 = (\pm i\boldsymbol{k}E_1 \pm \boldsymbol{i}\mathbf{p}_1),$$

but, by directly incorporating a mass term, we could use:

$$(\pm i\boldsymbol{k}E_1 \pm \boldsymbol{i}\mathbf{p}_1 + \boldsymbol{j}m_1).$$

So

$$\gamma^\mu(1-\gamma^5)p_1\gamma_\nu(1-\gamma^5)p_3$$
$$= (i\boldsymbol{k} + \mathbf{1}i)(1 - i\boldsymbol{j})(\pm i\boldsymbol{k}E_1 \pm \boldsymbol{i}\mathbf{p}_1)(i\boldsymbol{k} \ + \mathbf{1}\boldsymbol{i})(1 - i\boldsymbol{j})(\pm i\boldsymbol{k}E_2 \pm \boldsymbol{i}\mathbf{p}_2)$$
$$= (i\boldsymbol{k} + 1i - i - i1k)(\pm i\boldsymbol{k}E_1 \pm \boldsymbol{i}\mathbf{p}_1)(i\boldsymbol{k} + \mathbf{1}\boldsymbol{i} - \boldsymbol{i} - i\mathbf{1}\boldsymbol{k})(\pm i\boldsymbol{k}E_2 \pm \boldsymbol{i}\mathbf{p}_2).$$

Using

$$(i\boldsymbol{k})(\pm i\boldsymbol{k}E_1 \pm \boldsymbol{i}\mathbf{p}_1)(i\boldsymbol{k}) = (\pm i\boldsymbol{k}E_1 \mp \boldsymbol{i}\mathbf{p}_1)$$
$$(\mathbf{1}\boldsymbol{i})(\pm i\boldsymbol{k}E_1 \pm \boldsymbol{i}\mathbf{p}_1)(\mathbf{1}\boldsymbol{i}) = (\pm i\boldsymbol{k}E_1 \mp \boldsymbol{i}\mathbf{p}_1)$$
$$(-\boldsymbol{i})(\pm i\boldsymbol{k}E_1 \pm \boldsymbol{i}\mathbf{p}_1)(-\boldsymbol{i}) = (\pm i\boldsymbol{k}E_1 \mp \boldsymbol{i}\mathbf{p}_1)$$
$$(i\mathbf{1}\boldsymbol{k})(\pm i\boldsymbol{k}E_1 \pm \boldsymbol{i}\mathbf{p}_1)(i\mathbf{1}\boldsymbol{k}) = (\pm i\boldsymbol{k}E_1 \mp \boldsymbol{i}\mathbf{p}_1)$$

we obtain a total of $4(\pm i\boldsymbol{k}E_1 \mp \boldsymbol{i}\mathbf{p}_1)$ for this scalar product, or, for a state vector representing an antifermion (where $E_1 \to -E_1$), this would become $4(\mp i\boldsymbol{k}E_1 \mp \boldsymbol{i}\mathbf{p}_1)$.

For a vertex involving a fermion, with state vector $(\pm i\boldsymbol{k}E_2 \pm \boldsymbol{i}\mathbf{p}_2)$, taken over all four terms in the Dirac spinor,

$$4(\pm i\boldsymbol{k}E_1 \pm \boldsymbol{i}\mathbf{p}_1)(\pm i\boldsymbol{k}E_3 \pm \boldsymbol{i}\mathbf{p}_2) = 4 \times 4(E_1E_2 - \mathbf{p}_1\mathbf{p}_2) = -16(p_1 \cdot p_2).$$

The equivalent of $Tr[\gamma^\mu(1-\gamma^5)p_1\gamma^\nu(1-\gamma^5)p_3]\, Tr[\gamma_\mu(1-\gamma^5)p_2\gamma_\nu(1-\gamma^5)p_4]$ becomes 256 $(p_1 \cdot p_2)$ $(p_3 \cdot p_4)$, leading once again to a spin-averaged probability:

$$|\mathcal{M}|^2 = 64G^2(p_1 \cdot p_2)(p_3 \cdot p_4).$$

This approach is only valid for antifermion–fermion vertices, where the $V - A$ term $\gamma^\mu(1-\gamma^5)$ is included — that is, where the interaction is dipolar and single-handed. Otherwise, the product of the two scalar products does not correspond with the product of the two traces. In this method, however, the terms $p_1 \cdot p_2$ and $p_3 \cdot p_4$ can be easily extended to become scalar products of nilpotent operators where mass is to be taken into account.

7.5 BRST quantization

Though the Dirac nilpotent operator is automatically second quantized, and so already incorporates a full quantum field representation, it is interesting to look at more conventional approaches to field quantization. One of these helps us to demonstrate the relation between charge and energy operators, on which our formalism was constructed. In standard theory, field quantization requires gauge fixing (or the removal of gauge invariance) before propagators can be constructed. The canonical quantization of the electromagnetic field uses Coulomb gauge, but this means that Lorentz invariance must be broken. The path integral approach allows us to use any gauge, and so maintain Lorentz invariance, but the problem now is the introduction of nonphysical or 'fictitious' Fadeev–Popov ghost fields. A version used in string theory (BRST) eliminates the ghost fields by packaging all the information into a single operator, applied to the Lagrangian.

Significantly, the BRST operator (δ_{BRST}) is a nilpotent. This operator can be used to construct a Noether current (J_μ), corresponding to a nilpotent BRST conserved fermionic charge (Q_{BRST}). The condition for defining a physical state then becomes

$$Q_{\text{BRST}}|\Psi\rangle = 0.$$

In the Dirac nilpotent formulation, $(\pm i\boldsymbol{k}E \pm \boldsymbol{i}\mathbf{p} + \boldsymbol{j}m)$, which applies only to physical (mass shell) states, is already second quantized, and a nilpotent operator of the form δ_{BRST}. It is, also, a nilpotent *charge* operator of the form Q_{BRST}, but extended to incorporate weak and strong, as well as electromagnetic charges. It is, finally, in its eigenvalue form, identical to $|\Psi\rangle$. So the three possible meanings for the expression $(\pm i\boldsymbol{k}E \pm \boldsymbol{i}\mathbf{p} + \boldsymbol{j}m)$ apply respectively to: E and $\mathbf{p}$ interpreted as differential operators in time and space; E, $\mathbf{p}$ and m as coefficients determining the nature of the charges specified by $\boldsymbol{k}$, $\boldsymbol{i}$ and $\boldsymbol{j}$; and E and $\mathbf{p}$ interpreted as eigenvalues of energy and momentum. The nilpotent Dirac operator thus supplies simultaneously all the characteristics which the separate BRST terms δ_{BRST}, Q_{BRST}, and $|\Psi\rangle$ require.

7.6 Mass generation

If we consider the bosonic structures we have outlined as defining the vertices for boson production via the weak interaction, then it appears from the impossibility of creating a massless spin 0 boson that the two sharply defined helicity states for hypothetical massless fermions (Chapter 5) ensure that the pure weak interaction requires left-handed fermions and right-handed antifermions. In other words, it requires both a charge-conjugation violation and a simultaneous parity or time-reversal violation.

We can see in principle how this leads to mass generation by some process at least resembling the Higgs mechanism. Let us imagine a fermionic vacuum state with zero mass, say $(i\boldsymbol{k}E + \boldsymbol{i}\mathbf{p})$. An ideal vacuum would maintain exact and absolute C, P and T symmetries. Under C transformation, $(i\boldsymbol{k}E+\boldsymbol{i}\mathbf{p})$ would become $(-i\boldsymbol{k}E-\boldsymbol{i}\mathbf{p})$, with which it would be indistinguishable under normalization. No bosonic state is required for such a transformation, where the states are identical. If, however, the vacuum state is degenerate in some way under charge conjugation (as supposed in the weak interaction), then $(i\boldsymbol{k}E + \boldsymbol{i}\mathbf{p})$ will be transformable into a state which can be distinguished from it, and the bosonic state $(i\boldsymbol{k}E+\boldsymbol{i}\mathbf{p})(-i\boldsymbol{k}E-\boldsymbol{i}\mathbf{p})$ will necessarily exist, the $(i\boldsymbol{k}E+\boldsymbol{i}\mathbf{p})$ here being the creation of the new state, and the $(-i\boldsymbol{k}E - \boldsymbol{i}\mathbf{p})$ being the annihilation of the old state. This can only be true, however, if the state has nonzero mass and becomes the spin 0 'Higgs boson' $(i\boldsymbol{k}E + \boldsymbol{i}\mathbf{p} + \boldsymbol{j}m)(-i\boldsymbol{k}E - \boldsymbol{i}\mathbf{p} + \boldsymbol{j}m)$. The mechanism, which produces this state, and removes the masslessness of the boson, requires the fixing of a gauge for the weak interaction (a 'filled' weak vacuum), which manifests itself in the massive intermediate bosons, W and Z.

From the structures of bosons and the consideration of fermion spin, it would seem that mass and helicity are closely related. If the degree of left-handed helicity is determined by the ratio $(\pm)\boldsymbol{i}\mathbf{p}/(\pm)i\boldsymbol{k}E$, then the addition of a mass term will change this ratio. Similarly, a change in the helicity ratio will also affect the mass. If the weak interaction is only responsive to left-handed helicity states in fermions, then right-handed states will be intrinsically passive, so having no other function except to generate mass. The presence of two helicity states will be a signature of the presence of mass. The $SU(2)$ of weak isospin, which, in effect, expresses the invariability of the weak interaction to the addition of an opposite degree of helicity (due to the presence of, say, mass or electric charge) is thus related indirectly to the $SU(2)$ of spin, which is a simple description of the existence of two helicity states. It is significant that the *zitterbewegung* frequency, which is a measure of the switching of helicity states, depends only on the fermion's mass. Mass is in some sense created by it, or is in some sense an expression of it. The restructuring of space and time variation or energy and momentum, via the phase factor during an interaction, leads to a creation or annihilation of mass, which manifests itself in the restructuring of the *zitterbewegung*.

7.7 String theory

String theory was developed in an attempt to unify the four interactions and to remove the divergences assumed to be a consequence of defining point-like particles as the sources of physical interactions. Fundamental theorems suggest that 10 dimensions are the number required to remove all the anomalies from physics. The problem is to show why these 10 reduce to 4 dimensions of space-time in the world we actually observe, and the argument is that they are 'compactified'. The famous analogy with a hosepipe suggests that a compactification in this form, rolling the dimension in a circle around an uncompactified dimension, has connections with the Kaluza–Klein theory which extends Einstein's general relativity to 5 dimensions to include electric charge. The circle compactification also has the $U(1)$ symmetry required of the electric interaction. Whether the dimensionalities of string theory are all spatial or temporal dimensions becomes indeterminate if the compactification to the extra dimensions occurs below the Planck length, as is generally assumed.

String theory has never been intended to be a model-dependent theory, but this aspect is meant to be a stage in leading to a more abstract, general theory to which the models will be approximations. The objective is to find the correct vacuum which will generate the required compactification. There are famously many possible string theories, but they can be classified into 5 types, and a theory known as membrane theory or M-theory proposes that the different types can be combined if an extra dimension is added to connect them. The objects of M-theory are 2-dimensional spatial objects in 1-dimensional time, as opposed to the 1-dimensional spatial objects in 1-dimensional time of string theory. The combination of the 5 types of objects in 1-dimensional string theory in M-theory, however, has not yet happened. There is only an indication that it might be possible.

The main interest for us in string theory is not in producing a new one, but in the claim about the fundamental significance of 10 dimensions in removing anomalies, for it is immediately apparent that the nilpotent operator $(\pm i\boldsymbol{k}E \pm \boldsymbol{i}\mathbf{p} + \boldsymbol{j}m)$ can, in fact, be regarded as a 10-dimensional object embedded in Hilbert space or equivalent. The reason why it can be regarded as 10-dimensional originates in the fact that it expresses a fundamental duality between two 'spaces'. Though each of these is intrinsically 3-dimensional, each becomes expanded to five in the process of combination. Thus, there are five dimensions for iE, $\mathbf{p}$, m and 5 for $\boldsymbol{k}$, $\boldsymbol{i}$, $\boldsymbol{j}$; and six of these (all but iE and $\mathbf{p}$, or the equivalent time and space) are compactified in the sense of being fixed. The double 5-dimensionality also has connections with the Kaluza–Klein theory, which was originally two separate theories, one of which tried to explain invariant mass and the other electric charge, for the fifth term in the nilpotent has both these characteristics and it also represents a $U(1)$ symmetry.

Now, John Baez has claimed that the 10 dimensions indicate the existence of 2-dimensional strings (1 dimension of space, 1 of time) in an 8-dimensional space, which could be octonion, and has suggested that an octonion space is the true basis of physics.[12] Clifford algebra has the property of incorporating different dimensionalities in the same expression, and the nilpotent algebra is also an 8-dimensional object in two related senses. Two of the dimensions are made redundant because of the nilpotent structure, and the whole structure originates in the combination of the 4 dimensions of space and time and the 4 of mass and charge as a *broken* octonion (see Chapter 4).[13] Yet another way of looking at the fermionic nilpotent is as a 3-'dimensional' structure in $\boldsymbol{k}$, $\boldsymbol{i}$, $\boldsymbol{j}$, though point-like in space with one of the three 'dimensions' redundant, and this also corresponds to the 2-dimensional objects of string theory and the 3-dimensional ones of membrane theory. In this sense, the 2 or 3 redundant dimensions, which would, in string or M-theory be a string or brane connecting two 3- or 4-dimensional systems, become reduced to a point with zero dimension.

A classic prescription for a perfect string theory is one in which 'self-duality in phase space determines vacuum selection'.[14] The nilpotent certainly fulfils this criterion and it is also a mass-shell system and incorporates the right groups. It incorporates gravity-gauge theory correspondence and the holographic principle. (These are discussed in Chapter 8.) Though we have no need for a model-dependent theory to incorporate the interactions, it is important to be able to satisfy all the conditions that appear to make string theory, *or a more fundamental abstract theory, of which the model-dependent theories are approximations*, seemingly necessary. In a sense, the reduction of the 2-dimensional strings or 3-dimensional membranes to points in real space is the ultimate reduction of the model-dependent objects of string theory or M-theory to abstraction — a 'string theory without strings'. It is significant that the nilpotent formalism achieves this through solving the problem of vacuum.

In string theory, mass is generated by the vibrations of the strings, which replace point particles. However, this mass-generating mechanism is already incorporated in the point particle concept (as the Lamb shift makes clear), and relates to the Berry phase and *zitterbewegung*. It comes from the dual vector spaces needed to define a point particle, because the duality ensures that *zitterbewegung* (and hence vacuum fluctuation, the Lamb shift, *etc.*) is the origin of fermionic mass, and it *requires* a pole or singularity. In the nilpotent theory, rest mass always comes from defining a singularity through a double vector space. The very act of defining a point particle is also the same as ensuring that it undergoes vacuum fluctuations, or equivalent, and therefore generates mass. Again connecting with string theory, it is the same duality as that between gravity and gauge theory or between the local and nonlocal.

7.8 One-fermion theory

Of the various attempts to provide a 'physical' picture of quantum mechanics, one is especially interesting with reference to the nilpotent connection between fermion and vacuum. This is John Wheeler's 'one-electron theory of the universe', now extended to a 'single fermion theory'. Here, a single fermion in different spatial and temporal states becomes equivalent to many fermions appearing simultaneously. All the other fermions after the first are this fermion in different space and time states, positive and negative. In the nilpotent formalism, of course, each single fermion sees all other fermions as constituting a vacuum which is a mirror image of itself. It is relevant here that the usual objections against the one-fermion theory no longer apply, as the total fermionic structure requires equal numbers of fermions and antifermions existing simultaneously in *two different* but completely dual vector spaces — 'real' (or observed space) and vacuum space. In addition to exact equality between fermions and antifermions, the dual spaces also ensure that there is no mutual annihilation.

A consequence of this representation is that we can take an ensemble of fermions as a single fermion, and so justify applying the nilpotent condition in some form to larger structures, as is also evident from the fact that a version of the nilpotent Dirac equation using a discrete version of calculus (involving commutators rather than differentials) applies to classically discrete, as well as quantum, systems (see Chapter 5). The scale-independence of the single fermion/ensemble duality also fits exactly with the renormalization group procedure.

An ensemble is not localised as narrowly as a single fermion, so its vacuum will not be quite as nonlocal. If we take the whole universe to be a fermion, we can imagine all the possible space and time conditions (and bit flips in the terminology of Seth Lloyd[15]) as constituting this universe — which is what we have called vacuum. This includes all the states to which the single fermion could possibly aspire over time. In other words a real single fermion includes the entire possible history of the universe within its event horizon (the backward causality we referred to), making sense of our thermodynamics and evolutionary theory as a unique birth-ordering. However, this does not require determinism because we can only define the entire history if we localise the fermion exactly, which of course we cannot do. It is only an ideal. So, we have an exact idea of what we mean by nonlocal, as all the other potential states in space and time would of course be determined by the real states. The symmetry is perfect. Fixing a particular moment in time is localising in time, in the same way as we localise in space.

7.9 Dualities in nilpotent quantum theory

The nilpotent quantum theory is built upon fundamental dualities, and many dualities are intrinsic components of its structure. The operator and wavefunction are dual, but there are also dualities at many other levels. A related duality between fermion and vacuum originates from the principle that, by defining a fermionic state, we are also defining a fundamental singularity. To define a singularity we are forced to use a dualistic structure by simultaneously defining what is not singular. While we can view the fermion as a singularity with connections leading out to the rest of the universe, the vacuum acts as a kind of 'inverse singularity', with connections from the rest of the universe leading into the singularity constituting the fermionic state.

This duality ensures that vacuum is not something separated from the fermion. It is an intrinsic component of its definition, and of the spinor structure needed to define the fermion as a singular state. It is the reason why the fermion has half-integral spin — we can only define it by simultaneously splitting the universe into two halves which are mirror images of each other. The duality manifests itself physically in the phenomenon of *zitterbewegung*. Using either operator or amplitude, we define $(\pm i\boldsymbol{k}E \pm \boldsymbol{i}\mathbf{p} + \boldsymbol{j}m)$ as a 4-spinor, with 4 terms (each of which is nilpotent) arranged as a column/row vector. In the convention we have used, the 'real' state (the one subject to physical observation) is determined by the signs of E and $\mathbf{p}$ in the first term. The other three states are like three 'dimensions' of vacuum, the states into which the real term could transform by respective P, T or C transformations. The duality ensures that fermion and vacuum occupy separate 3-dimensional 'spaces', which are combined in the

γ algebra defining the singular state. It can be shown that these 'spaces', though seemingly different, are truly dual, each containing the same information, and that this duality manifests itself directly in many physical forms.

Some examples of the duality can be listed. The following effects have explanations based on real space (coded blue), and alternative explanations based on 'vacuum space' or, as I sometimes describe it for totality zero, 'antispace' (coded red):

Pauli exclusion	antisymmetric wavefunctions	nilpotency
angular momentum	p_1, p_2, p_3	iE, p, m
spin $\frac{1}{2}$	anticommutation of $\mathbf{p}$	Thomas precession
SR velocity addition	2 space components	space and time
holographic principle	space $\times$ space	space $\times$ time

In the case of Pauli exclusion, we can represent both p_1, p_2, p_3 and iE, p, m on orthogonal axes, to give a resultant 'vector' that is unique for each state. There are two mappings, on $\boldsymbol{\sigma}^1$, $\boldsymbol{\sigma}^2$, $\boldsymbol{\sigma}^3$ and on $\boldsymbol{\Sigma}^1$, $\boldsymbol{\Sigma}^2$, $\boldsymbol{\Sigma}^3$, and these are dual. Both sets of coordinates yield information about the same physical quantity: angular momentum. For full specification, angular momentum requires three separate pieces of information — magnitude, direction and handedness — and this is provided when iE, $\mathbf{p}$ and m are combined. It is also provided when we use all the information incorporated in the $\mathbf{p}$ vector alone.

We have already discussed how we can derive spin $\frac{1}{2}$ for fermions either by taking the commutator of the spin pseudovector $\sigma = -\mathbf{1}$ and the Hamiltonian, and deriving the half-integral spin from the anticommuting aspects of the components of $\mathbf{p}$, or, in terms of the Thomas precession, which is a relativistic correction. That is, we can derive spin $\frac{1}{2}$ using either the (multivariate) vector properties of space (using $\boldsymbol{\sigma}^1$, $\boldsymbol{\sigma}^2$, $\boldsymbol{\sigma}^3$) *or* the relativistic connection between space and time (using $\boldsymbol{\Sigma}^1$, $\boldsymbol{\Sigma}^2$, $\boldsymbol{\Sigma}^3$); the 3-dimensional 'spaces' involved are totally dual. The same applies to the velocity addition law in special relativity, which can be derived using either two dimensions of space (which generates the $\boldsymbol{\sigma}^1$, $\boldsymbol{\sigma}^2$, $\boldsymbol{\sigma}^3$ structure of Euclidean space) or one of space relativistically connected with one of time (which generates the $\boldsymbol{\Sigma}^1$, $\boldsymbol{\Sigma}^2$, $\boldsymbol{\Sigma}^3$ connection between space, time and proper time). The holographic principle provides another example where this occurs (and this is completely defined for the fermionic case by the nilpotent structure); here, the bounding 'area' specifying a system can be defined either by two spatial coordinates or one of space and one of time.

In all these examples, the two vector spaces required to define fermionic structure as a singularity are completely dual, though the symmetry of one is preserved while that of the other is broken. As in the parameter group from which it is derived, one condition is necessary to define the opposite in the other, and, in this way, the opposing conditions ultimately provide the same information in the same way as localised fermionic state and nonlocalised vacuum, or operator acting on phase factor and amplitude. *Zitterbewegung* is generally interpreted as a switching between a fermionic state and its vacuum, but it is also an expression of the duality between the 'real' space of $\boldsymbol{\sigma}^1$, $\boldsymbol{\sigma}^2$, $\boldsymbol{\sigma}^3$ and the 'vacuum space' of $\boldsymbol{\Sigma}^1$, $\boldsymbol{\Sigma}^2$, $\boldsymbol{\Sigma}^3$, neither of which is privileged. Both give an equally correct description of the state and must be simultaneously valid, even though we can only observe one at any given moment, and even the choice of broken/unbroken rotational symmetries between the components can be reversed by switching the space of observation from 'real' to 'vacuum' space.

The intrinsically dualistic nature of the fermion is most readily apparent when it is described by the self-dual nilpotent form of quantum mechanics, which is founded on the commutative combination of two vector spaces, each of which is exactly dual to the other. It is remarkable that physics has never been successfully founded on a concept of a single space, however distorted, but it does seem to respond well to being structured on two! From the initial duality, many others emerge, for example, those between fermion and vacuum, fermion and vacuum boson, operator and amplitude, nilpotent and idempotent, and broken and unbroken symmetries. These dualities allow the same mathematical structures (or the same structures but for sign changes) to describe apparently dissimilar objects, and so explain how the creation of a fermionic singularity effectively splits the universe into two halves that are mathematically and physically, if not observationally, equivalent.

Chapter 8

Gravity

8.1 General relativity or quantum mechanics?

Everyone recognises that gravity is one of the most difficult problems in physics. To try to make progress here, we will definitely have to go back to fundamental thinking. We are going to have to use the information available from Chapters 3 and 4 to make key decisions, and to use the methodology outlined in Chapter 2, which rejected compromise and model-dependent thinking in favour of abstractions and symmetries. If we believe what the fundamental symmetries tell us, we will have to accept what they imply, and see if the consequences match up to experiment.

At some point, any physicist thinking about fundamentals is going to have to confront the biggest dilemma facing us at the present time: quantum mechanics and general relativity cannot both be correct. One is going to have to be modified or abandoned. Perhaps both will. Every theory which tries to tackle this problem — for example, string theory, membrane theory, quantum loop gravity — makes such changes. What is changed depends on where you start. It looks like one or the other of these theories must be our starting point, because their results will have to be explained in the new theory, and no other viable starting point has ever been proposed. So, is general relativity the basis of fundamental physics or is it quantum mechanics? Interestingly, the bias so far has always been in the direction of starting from general relativity. For some reason, general relativity's appeal to something more tangible or accessible to human experience has prevailed over the abstract calculating engine of quantum mechanics. Nevertheless, I make no secret of the fact that, for me, the only route to a fundamental theory lies in quantum mechanics. There is no doubt that quantum mechanics has been put to far more extensive and rigorous testing than general relativity, and that it explains many more physical effects and in much greater detail and precision. General relativity has explained only a few things and these are at a considerable distance from its core principles.

So, can we accommodate general relativity if we privilege quantum mechanics? My answer is that we can, as long as we realise that it is primarily a *mathematical*, and not a physical theory, and that it is in no way an *explanation* of gravity, but a mathematical description of its effects on space and time as we *measure* them. This, I believe, is another area where our lack of knowledge of the true history has been damaging to our understanding of the physics. The problem is that the relation between Einstein's and Newton's theories has been determined by the historical verdict that one is a *revolutionary replacement* of the other, the physical description of gravity as a force being superseded by the geometrical description of a Riemannian space-time, in which the time component is indistinguishable from space, although this is in contradiction to its status in quantum mechanics. Such an argument, however, takes no notice of the fact that GR makes no physical sense on its own until it replaces its mathematically-derived curvature with the physically-derived Newtonian potential. Taken on their own, the GR field equations are simply a mathematical description of space-time curvature. It is only when they approach the 'Newtonian limit' that an arbitrary connection with a Newtonian 'physical' force can be established.

The Newtonian explanation uses an inverse-square force, a typical result of 3-dimensional spherical symmetry, analogous to the Coulomb force for charge and responding to the $U(1)$ symmetry group. If we consider this only as an empirical and experimental 'limit', we lose a possible fundamental symmetry with a fundamental meaning which we have to reintroduce arbitrarily as an 'approximation' which has no explanation. Since we are convinced that symmetries are more fundamental, in many respects, than mathematically sophisticated theories, we will have to make a decision on whether this symmetry is really broken in nature and, if it is, find some explanation why it appears to be 'approximately' true.

8.2 Gravity and quantum mechanics

Going back to quantum mechanics, we have to ask why gravity has never been successfully quantized. Quantum gravity presents us with nonrenormalizable infinities, largely because of the fact that its sources all have the same sign and there is no way it can be 'shielded', unlike any of the other forces. Quantum field theory attributes this to its intrinsically attractive force requiring a spin 2 mediating boson. Simple quantization principles also present us with the worst fit to data ever recorded in physics in deriving a value for the cosmological constant Λ. In fact, it is not only the worst ever fit to data, it is also the worst *possible* fit. A basic quantum-style argument makes Λ about 10^{120} to 10^{123} times higher than experimental data, based on 'dark energy', would suggest. Now, if, as Seth Lloyd has proposed, on the basis of the holographic principle, there are 10^{123} possible 'bit flips of data' in the observable universe over a Hubble lifetime,[16] then the prediction is as wrong as it could conceivably be. Not even 'not even wrong'!

A prediction that is so completely wrong is not so much a problem, more an opportunity. It seems to be telling us that we should be looking for an answer in entirely the opposite direction to the one we have assumed. Gravity is certainly quantum — it is almost inconceivable that anything involving matter could not be — but maybe 'quantum' needs to be taken in a wider context than the one we have assumed. We have assumed that 'quantum' refers only to *local* interactions. However, 'quantum' also demands nonlocality, and, at present, we have no mechanism for nonlocal correlations. Gravity is clearly quite different from all the recognised local forces. Its source, mass-energy, is completely continuous throughout the universe, whereas the other sources come from discrete localised points. In fact the continuity of mass-energy could indicate that it is even more unlike discrete sources than we have already imagined; for example, the Higgs field might well suggest that the amount of mass-energy is the same at every point in space, and that changes occur only in its method of manifestation because of the presence of charges of various kinds.[b]

Gravity is also extremely weak, only equalling the strength of the local forces between particles if summed up on a *universal* scale. Further, the gravitational energy between centres of mass is negative, compared to the positive energy between similar centres of charge, and negative energy, in fermionic states, is associated with vacuum rather than real space. Now a possible route forward at this point, and a way of incorporating gravity into a nilpotent structure, is to suggest that the total vacuum $-(\pm \boldsymbol{ik}E \pm \boldsymbol{i}\mathbf{p} + \boldsymbol{j}m)$, which is partitioned by the $\boldsymbol{k}$, $\boldsymbol{i}$, $\boldsymbol{j}$ operators, can be thought of as the negative energy continuous gravitational vacuum, which supplies

[b]If this is true, then we should not expect to find fermionic particles with rest energies greater than the energy of the Higgs field.

the mechanism for the instantaneous transmission of quantum correlation. According to this way of thinking, the gravitational vacuum has a special nature as a kind of 'sum' of all the others. Such a possibility is also hinted at by the $3 + 1$ nature of charge and mass within the parameter group, and was present in the original theories of particle structures based on it.

A similar concept of gravity/gauge theory correspondence has now emerged in string theory. Essentially, gravity and gauge theory (strong and electroweak) are dual. This is also evident in the way that the holographic principle privileges gravity to obtain information about the entire system. One can be used to provide information about the other. The fundamental duality is that of the nonlocal (gravity) and local (gauge theory), which tells us why gravity is so weak and why it is not obviously a quantized force. (Later in this chapter, we will approach gravitational quantization through the inertial reaction.) It also appears to tell us why the 'cosmological constant' is at the opposite end of the possible physical scale (in information terms) to the one worked out from quantum gravity.

Gravity/gauge theory correspondence appears to show that gravity acts as a dual to the combined gauge theories of the electric, strong and weak interactions, and duality in our language usually refers to negation and totality zero. The same seems also to apply in the continuous combined vacuum, which we have associated with gravity and the negative value of the fermionic nilpotent, $-(\pm i\boldsymbol{k}E \pm \boldsymbol{i}\mathbf{p} + \boldsymbol{j}m)$, compared to the discrete specific vacua created by the partitioning introduced by the quaternionic charge operators $\boldsymbol{k}$, $\boldsymbol{i}$, $\boldsymbol{j}$. What we are proposing here is that, rather than applying to a single local quantum state, gravity is a manifestation of the total vacuum, $-(\pm i\boldsymbol{k}E \pm \boldsymbol{i}\mathbf{p} + \boldsymbol{j}m)$, the sole source of nonlocal correlation, and applies to the whole universe at once. Gravity, it would seem, provides the most directly observable aspect of nonlocality. This is entirely the logic of all the symmetry arguments we have so far produced, and we have not yet found evidence that the fundamental symmetries are anything but absolutely true. Our methodology insists on us following this logic without exception.

According to this reasoning, unless something is communicated instantly, then we are unable to complete quantum mechanics — it isn't any of the known local interactions, and, if it isn't gravity, then it must be outside of what we have defined as physics. Of course, nonlocal correlations must create local effects. So, what are the local effects? It would seem here that the local manifestation associated with gravity must be the positive energy inertial local reaction, which is what we really observe and equate to the gravity of localised clumps of matter, and which, for a fermion, we could represent by the nilpotent structure $(\pm ikE \pm i\mathbf{p} + jm)$ itself. This force, however, is fictitious and repulsive, which means that it could be 'quantized' in the local sense, as a repulsive force, and so require a spin 1 gauge boson, rather than the spin 2 particle which creates the renormalization problems associated with quantum gravity.

The need for a spin 1 mediator and QED-like theory in 'quantum gravity' has been discussed in many of my publications (see for example Ref. 12). There, it has been suggested that the continuity of mass-energy, the filled vacuum, the Higgs field, and the need for instantaneous correlation between Dirac states, together with the fact that energy does not actually move (as opposed to the form of its realisation in connection with a discrete state), require an instantaneous gravitational force, which is undetectable by direct observation, and only ever observed through the c-dependent inertial reaction on discrete fermionic or bosonic states, which, being repulsive, requires a mediator of spin 1.

So, according to this way of thinking, gravity provides a nonlocal dual to the combined interactions, which has inertia as its local manifestation. While we can do the calculations for weak, strong and electric forces from the perspective of the local structure of the fermionic state, and work iteratively to determine the effect of the rest of the universe, with gravity we have to do the calculation universally or recursively to work back to the inertial effect on the quantized fermionic state. Gravity is the carrier of all the information separately available from the other three forces, and, in fact, of all their local manifestations, but it is their inverse rather than their summation, exactly as reflected in its negative energy, and we can even regard the attractive nature of the gravitational interaction between like particles as a *result* of its being a vacuum, rather than a local force, and therefore requiring negative energy. If this is a true description, gravity will be able to complete quantum mechanics in a way that the proposed (but so far unsuccessful) 'quantum gravity' cannot, because if localised 'quantum gravity' is valid, there will be no vacuum through which it can act.

This has some very interesting consequences, a number of which have been part of this theory from the beginning, but have now begun to appear in other areas of physics. The holographic principle is a conjecture which states that all the information in a system is contained in the bounding area. In the nilpotent structure, the E and $\mathbf{p}$ terms create the effective 'bounding area' (the mass term being redundant as additional information). This is an area in 'vacuum space', bounded by $\boldsymbol{k}$ and $\boldsymbol{i}$, but it is dual to an area in real space, and can be observed directly by reversing the roles of vectors (connected to space) and quaternions (connected to charge) in the nilpotent structure. The holographic information will then determine the nature of the system, including connected information about its inertial mass and charge structure. A related concept is quantum holography, which is like ordinary holography but not degenerate. It has now been unequivocally demonstrated in the case of 'quantum holographic encoding in a two-dimensional electron gas',[16] but probably has a much more general application. In the nilpotent formalism, we can accommodate the operation of quantum holography, by defining E and m as the phase and reference phase.

8.3 General relativity and Newtonian theory

Where, then, does this leave general relativity, which assumes that the gravitational force is governed by a Lorentzian metric, with maximum speed at the speed of light? To answer this we need to recall that general relativity, unlike Newtonian theory, with its 'God's eye view', is an *observer-centred* theory. All observation is local and we still have to accommodate this fact in our mathematical description of gravity as a nonlocal correlation. It may well be that, in resolving this problem, we can find an explanation for the otherwise arbitrary use of the Newtonian potential in general relativity.

Newtonian theory, as used by Newton, is very different from the way it is often portrayed. The key early problem that general relativity had to solve was the anomalous precession of the orbit of the planet Mercury. By the end of the nineteenth century, Le Verrier and others had calculated, that, from the perturbing effects of all known bodies in the solar system, the orbit of Mercury should precess at a rate of 531 seconds of arc per century, whereas the measured rate was 574 seconds of arc, leaving 43 seconds of arc unexplained. Now, Newton had first introduced perturbation theory in the *Principia*, but he didn't do it in the way people often think. To solve the orbit of a body perturbed by some external influence, he created a system of coordinates varying in space and time such that, *in that system*, the body would still describe an ellipse, sweeping out equal areas in equal times and obeying the conservation of angular momentum.

As an example he calculated the effect of one or two of the larger planets on Mercury's orbit at about 240 seconds of arc per century. He did this specifically to show that, at the level of accuracy possible to observational astronomers in his own time, there would be no observable effect. It was therefore more useful to him to solve the anomalies that could then be observed, like those of the lunar orbit. One thing that would have made even less sense for him to pursue was the fact that the velocity of light was finite, though he knew perfectly well that this would have an effect of its own on astronomical observation, for this was, after all, the precise way in which the finite speed of light was discovered by Roemer in 1676, and Newton used this astronomical anomaly for an optical test of his own in 1691. In addition, he believed that light would be deflected in a gravitational field.[17] He made no calculation, though others did later in the century.

The point of this is to show that, to extend Newtonian theory to accommodate astronomical results obtained at the precision available at the end of the nineteenth century requires more than basic Newtonian theory, but Newton was fully aware of the technique that had to be used: the creation of a *varying* coordinate system in which those effects which disturb the otherwise perfectly elliptical orbit of a two-body system can be accommodated by making the orbit perfectly elliptical in the new set of coordinates. In principle, this is still the technique which general relativity uses: find a geometry in which the effects which disturb an otherwise perfect path — in this case a straight line — are accommodated into the structure and then assume that the perfect path is followed in the new geometry. The idea of substituting geometry for physical effect is, in no sense, new with general relativity, only the idea that there is *only* geometry, not real physical effect; but even this is only an idealisation of the process, for all practical calculations from general relativity restore the Newtonian potential. In fact, it is impossible to tell from the mathematical structure alone whether there really is a physical effect or whether there is only geometry, or whether the distinction has any real meaning. Subsequent to the adoption of Einstein's theory, Cartan showed that standard Newtonian gravitational theory, without the velocity of light, could be represented by a geometrical structure very like Einstein's.[18]

8.4 The effect of observation on nonlocal gravity

If we assume that gravity is instantaneous and nonlocal, in line with the well-established instantaneous action of the static gravitational field, but physical *observation* is local, involving time-delayed luminal or subluminal interaction, we can no longer use the Lorentzian space-time of a local coordinate system in the description of gravity. So, what happens if we do so, as we must? The answer will be that we will create the equivalent of a noninertial frame for the gravitational system, with the resulting appearance of fictitious inertial forces, and a rotation of the coordinate system. We may recall that Newtonian theory already uses inertial forces as a standard way of dealing with mismatches between theory and observation, and that this is needed every time Newtonian theory is used, because, although Newtonian theory is defined only for inertial frames, *all* experimental observations use frames which are noninertial. So, in effect, we avoid the problem by assuming that the noninertial frames are actually inertial, and then add on the purely fictitious centrifugal and Coriolis forces to accommodate the noninertial effects.

We can, in fact, show that a gravitational system involving instantaneous interaction will require dynamic equations in the space-time of measurement of exactly the form required to predict the relativistic effect of planetary perihelion precession. It is easy to see that the main difference from pure Newtonian theory will be the equivalent of the gravitational deflection of light. In effect, the straight lines of observation, found using local forces limited by the speed of light (and mostly by direct electromagnetic signals), will become rotated in the same way as rays of light. Now, in the absence of the gravitational field, a ray of light would travel in a straight line according to the equation:

$$\left(\frac{dr}{dt}\right)^2 + r^2\left(\frac{d\phi}{dt}\right)^2 = 0.$$

This is also the Newtonian 'straight-line' or 'default' position for a body of unit mass in the absence of any deflecting force or source of dynamic energy, whereas the application of a Newtonian potential $-GM/r$ to deflect a body of unit mass from straight line motion, in a system of total mechanical energy E, requires a dynamic equation of the form:

$$\left(\frac{dr}{dt}\right)^2 + r^2\left(\frac{d\phi}{dt}\right)^2 = \frac{GM}{r} - E.$$

Now, in a gravitational field, with Newtonian potential $-GM/r$, the equation for the light ray, according to general relativity, becomes

$$\left(\frac{dr}{dt}\right)^2 + \left(1 - \frac{2GM}{rc^2}\right)r^2\left(\frac{d\phi}{dt}\right)^2 = 0,$$

while the relativistically-corrected dynamical equation from straight line motion, in a system of total mechanical energy E, would require a dynamic equation of the form:

$$\left(\frac{dr}{dt}\right)^2 + \left(1 - \frac{2GM}{rc^2}\right) r^2 \left(\frac{d\phi}{dt}\right)^2 = \frac{GM}{r} - E.$$

This is the equation which leads to relativistic perihelion precession, and we can see that the only difference it displays from the Newtonian equation is the term on the left-hand side equivalent to the light bending. The effect is as if the gravitational potential had rotated the local coordinate system based on Lorentzian space-time, by an amount which added the term $-(2GM/rc^2)\ (d\phi/dt)^2$ to the equation of motion, and this constitutes the 'inertial force' term (in the form of 'inertial energy') expressing the fact that the ordinary process of measurement provides a noninertial frame for a gravitational system. It can even be seen in terms of a curvature of space and time though as an effect, rather than source, of the gravitational field. No such 'rotation' would occur, of course, if it was possible to use some 'absolute' system of coordinates separate from the process of measurement.

Now, general relativity has never constituted an *explanation* of gravity. As it is constructed, it is purely a theory which says that space-time is curved, and that curvature constitutes what we call gravity. However, the mathematics of the general relativistic field equations includes no physical content, only an expression of the curvature. We have to put in the physical content when we solve for special cases, when terms which play similar roles to those in Newtonian theory are identified with them. So, within this context, we are perfectly entitled to find any physical description which corresponds to the equations. In effect, what we find is that general relativity is a theory of the *effects* of gravity, and particularly those relating to observation. It brings us no nearer to understanding what gravity is, but neatly packages and codifies the effects that gravity produces using a geometric system. In our terms, it is about epistemology or measurement, rather than ontology or the thing in itself. It doesn't distinguish between whether the gravitational influence travels at the speed of light or whether we just measure it that way.

It might be objected that the equation of light deflection, on which this discussion is based, can only be derived using the full field equations of general relativity, and that this proves, for example, that gravity must travel at the speed of light or less. In fact, this has long been known to be incorrect. The light deflection (and even the perihelion precession) can be derived purely from special relativity, as long as we include length contraction as well as time dilation. (The presence of both effects is obvious in the Schwarzschild solution for a spherically-symmetric point source.) Light, of course, is special relativistic, and the gravitational part of the calculation is only an application of gravitational energy, treated classically, to the special relativity of light. There is no assumption that gravity is special relativistic. In any case, the special nature of light even allows a *Newtonian* derivation of the effect. Eighteenth century Newtonian calculations were based on the idea that light, because of its great speed, would follow a hyperbolic orbit, with eccentricity $e \gg 1$, in being deflected by a large gravitating body. Twentieth century authors[19] have assumed that this would be equivalent to taking the potential energy equation for a body already in orbit, with velocity c and angle $0°$ at distance of closest approach R to the central body:

$$c^2 = \frac{GM}{R}(e+1),$$

from which we calculate the half-deflection angle as $\delta \approx 1/e \approx GM/Rc^2$, and the full-deflection as $2GM/Rc^2$, which is only half the measured value. The doubling can be considered as due to the addition of the length contraction effect to the time dilation caused by the presence of the potential energy. However, light's 'velocity' c when it is emitted from its source is nothing to do with any dynamical orbit. Dynamical characteristics are only acquired when it is drawn into an orbit, for which we need the kinetic energy equation

$$\frac{1}{2}c^2 = \frac{GM}{R}(e+1).$$

This gives the correct deflection, and was, in fact, the basis of Johann von Soldner's calculation of 1801, although Soldner failed to get the correct deflection by effectively only doing the half-deflection integration. In fact, energy considerations should lead to the same results for special relativity and classical physics, because special relativity is deliberately constructed to preserve the classical conservation of energy.

8.5 The aberration of space

We have interpreted general relativity in the only way in which we believe it can be compatible with quantum mechanics, and with our understanding of the nature of mass-energy, as derived both by symmetry and from observation. It is totally in keeping with GR as an expression of a purely abstract mathematical structure, realised as geometry. The most important result is that the theory is linearised because the space-time curvature is no longer the *source* of the gravitational field as well as its consequence. Other people might try alternative interpretations involving a real speed of c, but they have no support in the mathematical structure which constitutes the theory. They also lead to anomalous results, due to the assumed nonlinearities, such as wormholes, infinite gravitational collapse, the violation of conservation laws, unrenormalizable infinities, and the clearly unsustainable theory of 'quantum gravity'. Some of the suggestions might entertain the public interest in science fiction, but they don't involve science which can be imagined as verifiable by observation. Other theories might propose complex and usually model-dependent schemes of 'unification', but they will not be based on foundational thinking, and are unlikely to find experimental confirmation any time soon.

Every effect predicted by GR which has been observed, and every effect which has been predicted to be observable within a reasonable time frame, including 'gravitational waves' (here gravitational-inertial waves), will follow from the GR field equations as interpreted here. They will be gravitational in the sense that they are *caused* by gravity, but they will not be gravitational in the sense that they express the *nature* of gravity. A fuller description might require us to use a term like 'gravitation-inertia', since both concepts are involved.

Without the nonlinearities, the GR field equations will show much greater accuracy to many more orders of magnitude than if they are present. J0348 is a neutron star with a mass of two solar masses, only the second discovered after J1614-2230 in 2010. It was previously thought that the maximum mass for a neutron star was 1.5 solar masses, although the result has now been pushed to 2 or 3. J0348 has a white dwarf companion, with short orbital period of 2.46 hours. According to results now announced, the system produces gravitational waves, but not the anticipated extra ripples expected from nonlinearity.[20] Clearly, at the level of sensitivity revealed by the current measurements, nonlinear effects are not seen, and some nonlinear models are already ruled out. Further experiments will be needed in this area, but we have here a first indication that the GR equations are preserved to a much greater accuracy than some interpretations would have previously imagined, but that is certainly predicted by the interpretation proposed here.[21]

All the relativistic effects, in this interpretation, will be due to the action of gravity on the space-time of measurement, and could be described, by analogy, as the 'aberration of space'. We incorporate this aberration by adding on the required inertial force terms, and this becomes equivalent to using the GR field equations as the combined expression of gravitation and inertia (as Einstein actually required when he invoked the principle of equivalence). The inertial force correction terms are, typically, of magnitude $-3GMv^2/r^2c^2$ or $-3GM\alpha^2/r^4c^2$, and they are identical, in all respects, to the terms representing the gravitational effect on energy transmitted at the speed of light.

So, if, to a light-ray geodesic defined in the absence of a gravitational field, we introduce inertial forces exactly sufficient to cover the effects produced by a gravitational potential $-GM/r$, we will, *by definition*, obtain the standard new geodesic equation expected under these conditions with the extra terms now interpreted as being equivalent to the rotation of the space-time coordinates. Our calculations will produce exactly what we would expect to obtain from the general relativistic field equations. Under the most general conditions, we would recover the full theory of general relativity, though with the emphasis shifted from ontology (the nature of the gravitational field) to epistemology (the nature of measurement in a gravitational field). Quantum aspects, however, would now be available on the basis that they would be concerned with local inertial repulsion and spin 1 transmission, and so generate a renormalizable theory.

8.6 Gravitomagnetic effects

Many of the effects due to the presence of c in GR equations can be dealt with by assuming the existence of a 'gravitomagnetic' field analogous to the magnetic field of classical electromagnetic theory. This is, in fact, a relatively standard procedure in GR, although there are some extra consequences here of the fact that c is itself gravitationally affected. In fact, many of the results (like most of the standard consequences of GR) can be derived from special relativity, with additional 'GR' modifications arising from the use of both time dilation and length contraction. The assumption of linearity is assumed in standard calculations because the fields are relatively weak, but here we assume that it is fundamentally true in any case.

Kolbenstvedt has, in fact, derived the equations for the gravitomagnetic field from special relativity, using a kinematical argument, but has allowed only the effect of time dilation as special relativistic because it requires only the principle of equivalence and the Doppler effect.[22] He considers that the contraction of measuring rods requires general relativity or the 'curvature of space'. Nevertheless, as we have already seen from the case of gravitational light deflection, the length contraction, or some other doubling effect, is a necessary component of a full special relativistic or even classical treatment, and this is the key case because it is really an expression of the effect of the gravitational field on space-time geometry.

We will allow for this doubling effect, but otherwise follow Kolbentsvedt's procedure, and consider an object of mass M moving with velocity $\mathbf{u}$ in the positive x-direction in the frame of the laboratory, and a particle of mass m moving with velocity $\mathbf{v}$ under its gravitational influence. In the rest frame of mass M, the Lagrangian $\mathcal{L}_0$ of the mass m particle can be found from the variational principle:

$$\delta \int (-mds) = \delta \int \mathcal{L}_0 dt_0 = 0,$$

where

$$ds_0^2 = \gamma^{-2} dt_0^2 - \gamma^2 dr_0^2$$

is the line element in the rest frame, and

$$\gamma^2 = \left| -\frac{2GM}{rc^2} = \right| - \frac{2\phi}{c^2}.$$

Integrating for the rest frame,

$$\mathcal{L}_0 \, dt_0 = -m(\gamma^{-2} c^2 \, dt_0^2 - \gamma^2 \, dr_0^2)^{1/2}.$$

Then, transforming to the laboratory frame, and neglecting higher order terms, we obtain

$$\mathcal{L} dt = -m[(c^2 + 2\phi)(dt - udx)^2 - (1 - 2\phi)(dx - udt)^2 - dy^2 - dz^2]^{1/2}.$$

Once again neglecting the higher order terms, and dividing by dt, we obtain the Lagrangian

$$\mathcal{L} = - \left(c^2 - v^2 + 2\phi - 8\phi \frac{\mathbf{u} \cdot \mathbf{v}}{c^2} \right)^{1/2}.$$

When $u \ll v \ll c$, and the rest energy term mc^2 and higher order corrections are neglected, the series expansion approximates to

$$\mathcal{L} = \frac{1}{2} mv^2 - m\phi + 4m\phi \frac{\mathbf{u} \cdot \mathbf{v}}{c^2}.$$

This expression may be compared with the standard Lagrangian for a particle of mass m and charge q moving with speed $v \ll c$ in an electromagnetic field determined by scalar potential ϕ and vector potential $\mathbf{A}$:

$$\mathcal{L} = \frac{1}{2}mv^2 - q\phi + q\frac{\mathbf{A} \cdot \mathbf{v}}{c}.$$

The equations are clearly analogous, with $4\phi\mathbf{u}/c$ being the gravitational equivalent of the electromagnetic vector potential $\mathbf{A}$.

8.7 Maxwell's equations for gravitomagnetism

We can now extend the argument. The rotational analogue of the magnetic field term $\mathbf{B} = \nabla \times \mathbf{A}$ becomes the 'gravitomagnetic' field

$$\boldsymbol{\omega} c = \nabla \times \left(4\phi\frac{\mathbf{u}}{c}\right) = \left(4\frac{\mathbf{u}}{c}\right) \times (\nabla\phi) = 4\frac{\mathbf{u}}{c} \times \mathbf{g}.$$

We can then set out a series of equations of the Maxwell type:

$$\begin{aligned}
\nabla \cdot \mathbf{g} &= 4\pi G\rho \\
\nabla \cdot \boldsymbol{\omega} c &= 0 \\
\nabla \times \mathbf{g} &= -c\frac{\partial \boldsymbol{\omega}}{\partial t} \\
\nabla \times \boldsymbol{\omega} c &= 4\pi G\rho\mathbf{v} + c\frac{\partial \mathbf{g}}{\partial t}.
\end{aligned}$$

Here, $G\rho\mathbf{v}$ takes the place of $\mathbf{j}$; and $4\pi G\rho$ that of ρ/ε_0. In empty space $\mathbf{g}$ and $\boldsymbol{\omega} c$ will satisfy the equations for $\nabla \times \nabla \times \mathbf{g}$ and $\nabla \times \nabla \times \boldsymbol{\omega} c$:

$$\begin{aligned}
\Box\mathbf{g} &= 0 \\
\Box\boldsymbol{\omega} c &= 0,
\end{aligned}$$

with the analogous mass density terms being added where sources are present. Other standard results follow immediately, as they do in electromagnetic theory: the Poynting vector (energy flux in an element of solid angle), energy density of fields, Lorentz force, Larmor precession frequency, equation of motion of a particle, quadrupole radiation, etc. Kolbentsvedt points out that such effects as "geodesic deviation of spinning particles, precession of gyroscopes orbiting the Earth, and 'dragging' of inertial frames by rotating masses, by leaning on well-known effects from classical electromagnetism and atomic physics involving spin–orbit and spin–spin coupling" would follow automatically from a gravitomagnetic theory.

The gravitomagnetic equations here are concerned with inertial effects and so repulsive forces, rather than attractive, which they would be if they were directly gravitational; this makes the rotational term ωc positive, unlike the gravitational field, and so, for comparison with Maxwell's equations, in which **E** and **B** are both positive, we take **g**, a 'static' component of inertial repulsion, in place of the gravitational field $-\mathbf{g}$ in the equations from which the wave solutions are obtained. The form of the equation $\nabla \cdot \mathbf{g} = 4\pi G\rho$ is, of course, insensitive to the velocity at which the interaction is transmitted, so a real gravitational attraction of the mass density ρ transmitted at infinite velocity would be formally indistinguishable from a fictitious static inertial repulsion *by* the mass density ρ transmitted at the velocity of light. We will show later that the static component of inertial repulsion, which is assumed by the principle of equivalence to be numerically equal to the attractive force of gravity, has a very important cosmological significance.

The factor 4 in the vector potential is the main other difference from the equivalent term in electromagnetic theory, and, as we have seen, it comes from a combination of the effects of length contraction and time dilation, which means that the pure term c is no longer the defining one as it is in electromagnetic theory. It emerges also in the theory of 'gravitational waves', as produced by other authors from general relativity, but it is not an indication of the need for a spin 2 boson in quantum gravity, as has been claimed, for its origin in special relativity is clear, and it is possible to assign this origin to a gravitational effect on space-time which has no reciprocal effect on the gravitational force itself.

8.8 Mach's principle

One of the things that general relativity never succeeded in resolving was the relation between gravitation and inertia. In particular, it was never able to accommodate Mach's principle, or the idea that the inertial mass of any object was due to its interactions with the rest of the matter in the universe. An argument that sought to overcome this problem, on the basis of gravitomagnetism, was put forward by Sciama in 1953,[23] though he later abandoned it. However, if we extend this argument on the basis of our new understanding of how gravitomagnetism and inertia are related, we find that it leads to a remarkable prediction which now appears to have a close relation to some significant experimental results discovered subsequently.

According to our understanding, relativistic and 'gravitomagnetic' effects emerge as properties of the local coordinate system and not as intrinsic properties of the gravitating source. If we are able to construct a set of Maxwell equations for the gravitomagnetic field that are exactly analogous to those used in electromagnetic theory, then we would expect gravitomagnetic analogues to all the effects that are known from electromagnetic theory. One such analogue would be an acceleration-dependent inductive force with the same structure as that appearing in electromagnetic theory. That is,

$$F = \frac{G}{c^2 r} m_1 m_2 \sin\theta \frac{dv}{dt}$$

would be the analogue of

$$F = \frac{q_1 q_2 \sin\theta}{4\pi\varepsilon_0 c^2 r} \frac{dv}{dt},$$

which can be derived from Faraday's law of induction.

This was exactly the force that Sciama proposed for explaining inertia. According to Sciama, the inertia of a body of mass $m = m_1$ might be derived from the action of the total mass $m_u = m_2$ within a Lorentzian event horizon of radius r_u, thus making the force F equal to Kma, with K a constant. Sciama required

$$F = \frac{Gm_1 m_2}{c^2 r} \frac{dv}{dt} \propto m_1 \frac{dv}{dt} = Kma.$$

So, if $m_2 = m_u$, the Hubble mass, is responsible for the inertia of a body of mass $m = m_1$, then $F = Kma$ if $Gm_u \propto c^2 r$, and $K = \frac{1}{2}$ if the zero total energy means that $Gm_u = \frac{1}{2} c^2 r$.

However, in our interpretation, the continuous and zero-gradient mass-field, or vacuum which provides gravitational nonlocality (say, the Higgs field), must produce a local consequence, analogous to unit charge, and define the standard for a unit inertial mass for the entire universe in exactly the same way as the almost constant gravitational field **g** allows us to define a unit mass at the Earth's surface. As a local effect, inertia, unlike gravity, is time-delayed, and limited by the velocity of light, and therefore also by the Hubble radius which defines its event horizon. So, the Hubble mass m_u will define a radial inertial field of constant magnitude from the centre of any given local coordinate system to the event horizon defined by r_u. At the same time, the gravitational field (Gm_u/r_u), independently of the local coordinate system, will define a unit of gravitational mass within the same radius. Supposing that isotropy removes the angular dependence, we apply the principle of equivalence and obtain:

$$\frac{Gm_u}{c^2 r}\frac{dv}{dt} = \frac{Gm_u}{r_u^2}.$$

We are saying that the object defined by mass m requires an inertial force, with an acceleration

$$a = \frac{dv}{dt} = v\frac{dv}{dr} = \frac{c^2 r}{r_u^2}$$

and, by integration with respect to r, a velocity related to the Hubble constant, H_0:

$$v = \frac{cr}{r_u} = H_0 r.$$

This is Hubble's law for cosmological redshift, which we have derived without assuming a cosmology to explain it, but we have also found something new, an acceleration term, which, in terms of Hubble's constant, can be written:

$$a = \frac{v^2}{r} = H_0^2 r.$$

If gravity is nonlocal, this acceleration will be 'fictitious', in the sense that it arises from using a local Lorentzian coordinate system to model an instantaneous interaction. It is a manifestation of a fundamental physical process, analogous to a centrifugal force, which does not depend on a prior assumption of cosmology of any kind, though it may *determine* it. Ultimately, fundamental physical effects require fundamental physical explanations, and one purely based on a historical cosmology — of things that just happened to happen at particular times — is not an adequate explanation in these terms. Cosmology, wherever possible, should follow physics, not the other way round.

If we now combine this repulsive force with the attractive gravitational force due to total mass m at any distance, we can calculate an equivalent vacuum density, ρ, from

$$F = \frac{Gm}{r^2} - H_0^2 r = \left(\frac{4}{3}\pi G\rho - H_0^2\right) r.$$

We can re-express this as a Poisson–Laplace equation in which

$$\nabla^2\phi = 4\pi G\left(\rho - \frac{3H_0^2}{4\pi G}\right) = 4\pi G\left(\rho + \frac{3P}{c^2}\right) = 4\pi G\left(\rho - 3\rho_{\text{vac}}\right).$$

We can identify here a vacuum density

$$\rho_{\text{vac}} = \frac{H_0^2}{4\pi G},$$

which is equivalent to a 'dark' energy density or negative pressure

$$-P = \frac{H_0^2 c^2}{4\pi G},$$

which can also be expressed through the cosmological constant

$$\lambda = 8\pi G\rho_{\text{vac}} = 2H_0^2.$$

If we define the critical density for a 'flat' universe as

$$\rho_{\text{crit}} = \frac{3H_0^2}{8\pi G}$$

we obtain

$$\frac{\rho_{\text{vac}}}{\rho_{\text{crit}}} = \frac{2}{3}.$$

What this means is that we have derived a vacuum density which is 67% of the density of the universe, assuming that this is at the critical value, and, of course, assuming uniform density and isotropy. It manifests itself through an outward inertial acceleration to the velocity generated in Hubble's law. The force law obeys Hooke's law, exactly as would be expected for a term equivalent to the cosmological constant.

This would be interesting just as explanation, but in fact it was originally a *prediction*. Many years after this prediction was first made, and published on at least four separate occasions (the last and most accessible of which was in a book called *A Revolution Too Far*, 1994[24]), an effect of exactly this kind was discovered. The latest data from the Planck satellite say that it is in the region of 68% of the energy of a universe at critical density.[25] It is described as the *dark energy*, and, according to all accounts, the effect was totally unexpected when it was first found and is still unexplained.

We have now seen that, if we equate the inertial reaction numerically with the undetectable gravitational attraction (so defining an equivalence principle), we justify a form of Mach's principle, and obtain gravitomagnetic effects, redshift, and acceleration of the redshift. In the simplest case of 'curvature', provided by a point source, we generate the Schwarzschild metric and a factor 4 in the gravitomagnetic equations by comparison with those for QED, incorporating a factor 2 for space 'contraction' and another 2 for time 'dilation'.

8.9 Can we quantize gravitational inertia?

The theories concerning the gravitomagnetic field and dark energy are interesting as showing a new aspect of fundamental reasoning — that it can cast theories created for quite different purposes into new forms, incorporating fundamental ideas not present in the original theories, and seemingly emerge with quite new results. The basic ideas of gravitomagnetism, the aberration of space, and the inertial origin of redshift with a concept of acceleration were known to me from fundamental reasoning before I knew of the theories of Kolbentsvedt and Sciama, but these theories created a much more powerful basis on which they could be constructed, leading to a much more exact and predictive treatment of the phenomena. Another example was a theory of discrete gravity based on 'extended causality' with a specific use of the proper time, which had the exact structure required to be developed using the nilpotent method of quantization. This is rather more technical than some of the other arguments in this chapter, and much of this development is due to the original authors, but, because it is such an important subject, it might just be mentioned as a possible means to bring to fruition the idea that the realisation of the local aspect of gravity in an inertial theory might be capable of quantization, as predicted.

According to our earlier arguments, the local theory which we have described as inertia should be susceptible to quantization. The interpretation of inertia as the result of the interaction between discrete matter and the continuous gravitational vacuum suggests that it is gravitational inertia rather than gravity which is subject to quantization. Quantizing a pure 4-dimensional structure has proved to be a problem, because time is not an observable in quantum mechanics as it is in classical relativity theory, nor is quantum space-time truly 4-dimensional, as the two parameters are added together only with the mediation of another structure based on the γ matrices or $\boldsymbol{i}$, $\boldsymbol{j}$, $\boldsymbol{k}$. If we assume that 3-dimensionality is the sole source for discreteness in physics, the mathematical object called a 4-vector will have no physical realisation at the quantum level. This means that, for a fully quantized theory, we need a metric other than the 4×4 representation using x, y, z, t, with added curvature, which is used in classical general relativity.

The obvious one that suggests itself is a 3×3 representation, with diagonal terms $\boldsymbol{i}\boldsymbol{k}t$, $\boldsymbol{i}\mathbf{r}$, $\boldsymbol{j}\tau$ in the absence of the curvature resulting from a gravitational field. This formalism would have the distinct advantage of being a natural $2 + 1$ theory of gravitational effects (the 2 representing the 'real' terms $\mathbf{r}$ and τ, and the 1 the imaginary term *it*, in interesting analogy to the $2 + 1$ structure of branes in M-theory), and such theories are already known to be renormalizable, unlike those with a higher number of dimensions. Instead of using a 4-dimensional structure, we use the 3-*dimensional* nilpotent structure, represented by the terms $\pm \boldsymbol{i}\boldsymbol{k}t \pm \boldsymbol{i}\mathbf{r} + \boldsymbol{j}\tau$ and $\pm \boldsymbol{i}\boldsymbol{k}E \pm \boldsymbol{i}\mathbf{p} + \boldsymbol{j}m$. The first term may be described as the 'quantum metric'; the second is the realisation of the Dirac state, and may be regarded as the phase space version of the first. No other fundamental structure is both fully quantum and fully relativistic, and the 3-dimensionality of the structure is essential to both of these conditions.

Significantly, the discrete theory also dispenses with the transverse directions to create a $1 + 1$ space-time, paralleling the fact that a quantum Dirac particle with conserved charge (the kind of object to which quantum gravity or quantum gravitational inertia will apply) requires only $\mathbf{r}$, and a single well-defined direction of spin, rather than the classical x, y, z. The 3-dimensionality comes from the fact that for gravitational inertial interactions involving individual fermionic states at the quantum level, there is an effective reduction or compactification of the spatial dimensions to a single well-defined parameter ($\mathbf{r}$). In this case, we can construct a $2 + 1$ theory of gravitational inertia based on a 3×3 'quantum metric' (with $\mathbf{r}$ and τ representing the 'real' parts and *it* the imaginary), which would be both quantizable and renormalizable. It would provide the full Dirac 'atom' solution and $U(1)$ QED-type behaviour with a corresponding photon-like mediating boson merely on the assumption of spherical symmetry and the multivariate vector nature of the spin term $\mathbf{p}$ (or, equivalently, conserved charge).

The standard mathematical representation of the gravitational force incorporates no information relating to speed, but the description of gravity as an undetectable property of the vacuum would *require* it to be instantaneous. The c-dependence of the inertial reaction, however, determines that, though linear and renormalizable, this force will itself be affected by gravity, giving rise to the 'curvature' terms in the metric tensor, as in general relativity. It is, however, the 'curvature' of the metric for inertia, not for gravity, which has no metric.

8.10 Extended causality and quantized inertia

A discrete theory of 'extended causality' has been proposed by de Souza and Silveira,[26] based on the fact that a single object (particle or field) at two points in Minkowski space-time (represented by the 4-vector x) must satisfy the causality constraint

$$\Delta\tau^2 + \Delta x^2 = 0$$

which defines a hypercone for the object. As soon as we convert this to the nilpotent form

$$\Delta(i\boldsymbol{k}t)^2 + \Delta(\boldsymbol{i}\mathbf{r})^2 + \Delta(\boldsymbol{j}\tau)^2 = 0$$

we are on the way to quantization, and through the multivariate nature of $\mathbf{r}$, it will automatically incorporate spin. Extended causality then applies when we shift τ and x by infinitesimal steps $d\tau$ and dx. We then obtain

$$\begin{aligned}(\Delta\tau + d\tau)^2 &+ (\Delta x + dx)^2 \\ &= (i\boldsymbol{k}\Delta t + i\boldsymbol{k}\,dt)^2 + (\boldsymbol{i}\Delta\mathbf{r} + \boldsymbol{i}d\mathbf{r})^2 + (\boldsymbol{j}\Delta\tau + \boldsymbol{j}\,d\tau)^2 = 0.\end{aligned}$$

Defining f as a fibre in the 'space-time' $x \equiv (\boldsymbol{i}\mathbf{r}, i\boldsymbol{k}t)$, where $f_\mu = dx_\mu/d\tau = \boldsymbol{i}d\mathbf{r}/\boldsymbol{j}d\tau + i\boldsymbol{k}\,dt/\boldsymbol{j}d\tau$ and $f^\mu = -d\tau/dx^\mu = -\boldsymbol{k}d\tau/d\mathbf{r} + i\boldsymbol{i}d\tau/dt$, we may combine these two equations to obtain:

$$\Delta\tau + f\cdot\Delta x = \boldsymbol{j}\Delta\tau + (\boldsymbol{i}d\mathbf{r}/\boldsymbol{j}d\tau + i\boldsymbol{k}\,dt/\boldsymbol{j}d\tau)(\boldsymbol{i}\Delta\mathbf{r} + i\boldsymbol{k}\Delta t) = 0,$$

while extended causality now requires

$$\begin{aligned}(\tau - \tau_0) &+ f_\mu(x - x_0)^\mu \\ &= \boldsymbol{j}(\tau - \tau_0) + (\boldsymbol{i}d\mathbf{r}/\boldsymbol{j}d\tau + i\boldsymbol{k}\,dt/\boldsymbol{j}d\tau)(\boldsymbol{i}\mathbf{r} + i\boldsymbol{k}t - \boldsymbol{i}\mathbf{r}_0 - i\boldsymbol{k}t_0) = 0.\end{aligned}$$

Suppose now that we introduce a massless scalar field $\phi_f(x,\tau) \equiv \phi_f(\boldsymbol{i}\mathbf{r}, i\boldsymbol{k}t, \boldsymbol{j}\tau)$. The extended causality will constrain it to the hypercone generator. The previous equation will also induce a direction to the field derivatives, so that

$$\partial_\mu\phi_f = (\partial_\mu - f_\mu\partial_\tau)\phi_f \equiv \nabla_\mu\phi_f.$$

With this expression we can now derive a discrete field equation. If χ is the coupling constant and $\rho(x,\tau) \equiv \rho(\boldsymbol{i}\mathbf{r}, i\boldsymbol{k}t, \boldsymbol{j}\tau)$ the discrete field source, then, using standard methods, the action of the field is given by

$$S_f = \int i d\mathbf{r} dt d\tau \left\{ \frac{1}{2} \eta^{\mu\nu} \nabla_\mu \phi_f \nabla_\nu \phi_f - \chi \phi_f \rho(\boldsymbol{i}\mathbf{r}, \boldsymbol{i}\boldsymbol{k}t, \boldsymbol{j}\tau) \right\}.$$

The field equation then becomes

$$\eta^{\mu\nu} \nabla_\mu \nabla_\nu \phi_f (\boldsymbol{i}\mathbf{r}, \boldsymbol{i}\boldsymbol{k}t, \boldsymbol{j}\tau) = \rho(\ \boldsymbol{i}\mathbf{r}, \boldsymbol{i}\boldsymbol{k}t, \boldsymbol{j}\tau),$$

with energy tensor

$$T_f^{\mu\nu} = \nabla^\mu \phi_f \nabla^\nu \phi_f - \frac{1}{2} \eta^{\mu\nu} \nabla^\alpha \phi_f \nabla_\alpha \phi_f.$$

The solution can be conveniently expressed in terms of a Green's function. Here we write:

$$\phi_f(\boldsymbol{i}\mathbf{r}, \boldsymbol{i}\boldsymbol{k}t, \boldsymbol{j}\tau) = \int i d\mathbf{r} dt d\tau (\boldsymbol{i}\mathbf{r}' + \boldsymbol{i}\boldsymbol{k}t') G_f(\boldsymbol{i}\mathbf{r} + \boldsymbol{i}\boldsymbol{k}t + \boldsymbol{j}\tau - \boldsymbol{i}\mathbf{r}' - \boldsymbol{i}\boldsymbol{k}t' - \boldsymbol{j}\tau') \times \rho(\boldsymbol{i}\mathbf{r}', \boldsymbol{i}\boldsymbol{k}t', \boldsymbol{j}\tau')$$

and

$$\eta^{\mu\nu} \nabla^\mu \phi_f \nabla^\nu \phi_f G_f(\boldsymbol{i}\mathbf{r} + \boldsymbol{i}\boldsymbol{k}t + \boldsymbol{j}\tau) = i\delta\mathbf{r}\delta t\delta\tau.$$

Using the Heaviside step function, Θ, with $b = \pm 1$, the Green's function is then

$$G_f(x, \tau) = \frac{1}{2} \Theta(b f^4 t) \Theta(b\tau) \delta(\tau + f \cdot x)$$

or

$$G_f(\boldsymbol{i}\mathbf{r} + \boldsymbol{i}\boldsymbol{k}t + \boldsymbol{j}\tau) = \frac{1}{2} \Theta(b(\boldsymbol{i}d\mathbf{r}/\boldsymbol{j}d\tau + \boldsymbol{i}\boldsymbol{k}\, dt/\boldsymbol{j}d\tau)\boldsymbol{i}\boldsymbol{k}t) \times \Theta(b\boldsymbol{j}\tau)\delta(\boldsymbol{j}\tau + (\boldsymbol{i}d\mathbf{r}/\boldsymbol{j}d\tau + \boldsymbol{i}\boldsymbol{k}\, dt/\boldsymbol{j}d\tau)(\boldsymbol{i}\mathbf{r} + \boldsymbol{i}\boldsymbol{k}t)).$$

Even in classical discrete form, this equation is independent of the transverse components of the field, paralleling the quantum reduction to a single well-defined direction of spin.

If we now take a single scalar charge $q(\tau)$, with world line $z(\tau)$, as a field source, then:

$$\rho(x, t_x = t_z) = q(\tau_z) \delta^{(3)}(x - z(\tau_z)) \delta(\tau_x - \tau_z),$$

and the solution for the emitted field becomes:

$$\phi_f(x, t_x) = \chi \int d\tau_y \Theta(t_x - t_y) \Theta(\tau_x - \tau_y) \delta[t_x - t_y + f \cdot (x - y)] q(\tau_z),$$

which reduces to

$$\phi_f(\boldsymbol{i}\mathbf{r}, \boldsymbol{i}\boldsymbol{k}t, \boldsymbol{j}\tau) = \chi q(\tau) \Theta(t) \Theta(\tau)|_f,$$

or

$$\phi_f(\boldsymbol{i}\mathbf{r}, \boldsymbol{i}\boldsymbol{k}t, \boldsymbol{j}\tau) = \chi q(\tau)|_f,$$

when $\tau \geq 0$ and $t > 0$. We have found that applying a massless scalar gives us a discrete field equation, and a field source represented by a scalar charge to generate a 'graviton'-like object and a metric for a discrete field related to gravity, though now the field is a repulsive inertial one, fully quantized via the Dirac nilpotent, and the 'graviton'-like object is correspondingly identified as a spin 1 boson or photon-like pseudo-boson. The field can only be quantized in this way because it is inertial, with positive energy, not gravitational, with negative energy, as negative energy, in our understanding, represents nonlocal vacuum rather than the local quantized state.

In the discrete model, the emission or absorption of a field causes a discrete change in q, and we can apply the standard techniques appropriate to quantum field theory, and, in particular, QED, to develop the formalism for interactions at higher orders, the $2+1$ nature of the theory ensuring its renormalizability. In effect, though with more difficulty, we have reversed the argument used for the other forces, working from vacuum to the quantized local state.

Chapter 9

Particles

9.1 Particle structures from nilpotent quantum mechanics

Particle (i.e. fermion) structures have a simple origin in fundamental terms. They are a description of point charges, electric, strong and weak. Essentially, all charges may be present or absent, which we can symbolise by 1 or 0. They may be + or −. They may also be e, s or w, and, if so, will acquire the characteristics with which these are associated when they are packaged together in the fermionic nilpotent, respectively scalar, vector and pseudoscalar. There are also considerations related to CPT symmetry and the chirality of fermions under the weak interaction. The different possible combinations of structures within these rules are what make the particle structures that are possible. Everything else follows automatically, except the particle masses, though even some aspects of this difficult subject are consequences of the same rules. Most of what can be achieved is via pattern recognition on a large scale, and there are a number of different ways of creating the patterns. Interestingly, though they have quite different starting points, they all seem to be pointing in the same direction. As in several other areas, an appeal to fundamental ideas often leads to a particular twist on well-known results and entirely new consequences. We will show a calculation that, reinterpreted on fundamental grounds, leads to a very important consequence with directly testable predictions.

Ultimately, the charge structures, like the fermions themselves, involve some kind of combination of two 3-dimensionalities, one with symmetry broken and the other with symmetry preserved (see Chapter 3). This is, as we have previously shown, how the point structure is created. One way of doing this is to go straight to the nilpotent representation, in which the four components of the spinor represent the full potentiality of what any fermionic state could be transformed into, with the weak, strong and electric interactions as the means of making this transfer. According to the ideas postulated in the last chapter, the gravitational or inertial interaction is 'passive' in this respect, the vacuum reflection (expressible as 1ψ or scalar $\times\ \psi$) leading to the state itself. ψ is taken to be the local 'inertial' manifestation of the fermion, with $-\psi$ the nonlocal gravitational dual. Now, of the two types of fundamental fermion, only quarks incorporate the explicit vector behaviour (i.e. showing a structure made of 3 components) of the momentum operator in their spinor state vectors. That is, in our postulated nilpotent formalism for a baryon (Chapter 6), each of the three nilpotents used to construct this state, which we now call (valence) quarks, only contains, at any instant, one component of the total momentum vector $\mathbf{p} = (\mathbf{p}_1, \mathbf{p}_2, \mathbf{p}_3)$ of the baryon, and that, in the allowed *phases* of the interaction, the total momentum is in *just one* of these components. However, as we saw in Chapter 2, we can create a nilpotent without specifying that $\mathbf{p}$ is a vector divisible into components, and so can construct a nilpotent fermion without having an initial separation into component parts. The result is an overall final structure which looks the same, but the first process has more simultaneous options than the second. We can account for the distinction in terms of the vector phase, which is here an instantaneous choice of direction, remembering that, as we have shown in Chapter 5, all the information about a fermion is contained in the direction of its momentum or spin vector. So a baryon state vector might have a form such as

$$\begin{matrix} \textit{inertial} \\ \textit{strong} \\ \textit{weak} \\ \textit{electric} \end{matrix} \begin{pmatrix} i\mathbf{k}E \pm i\boldsymbol{\sigma}\cdot\mathbf{p}_1 + \mathbf{j}m \\ i\mathbf{k}E \mp i\boldsymbol{\sigma}\cdot\mathbf{p}_1 + \mathbf{j}m \\ -i\mathbf{k}E \pm i\boldsymbol{\sigma}\cdot\mathbf{p}_1 + \mathbf{j}m \\ -i\mathbf{k}E \mp i\boldsymbol{\sigma}\cdot\mathbf{p}_3 + \mathbf{j}m \end{pmatrix} \begin{pmatrix} i\mathbf{k}E \pm i\boldsymbol{\sigma}\cdot\mathbf{p}_2 + \mathbf{j}m \\ i\mathbf{k}E \mp i\boldsymbol{\sigma}\cdot\mathbf{p}_2 + \mathbf{j}m \\ -i\mathbf{k}E \pm i\boldsymbol{\sigma}\cdot\mathbf{p}_3 + \mathbf{j}m \\ -i\mathbf{k}E \mp i\boldsymbol{\sigma}\cdot\mathbf{p}_2 + \mathbf{j}m \end{pmatrix} \begin{pmatrix} i\mathbf{k}E \pm i\boldsymbol{\sigma}\cdot\mathbf{p}_3 + \mathbf{j}m \\ i\mathbf{k}E \mp i\boldsymbol{\sigma}\cdot\mathbf{p}_3 + \mathbf{j}m \\ -i\mathbf{k}E \pm i\boldsymbol{\sigma}\cdot\mathbf{p}_2 + \mathbf{j}m \\ -i\mathbf{k}E \mp i\boldsymbol{\sigma}\cdot\mathbf{p}_1 + \mathbf{j}m \end{pmatrix}$$

or

$$\begin{matrix} \textit{inertial} \\ \textit{strong} \\ \textit{weak} \\ \textit{electric} \end{matrix} \begin{pmatrix} i\mathbf{k}E \pm i\boldsymbol{\sigma}\cdot\mathbf{p}_1 + \mathbf{j}m \\ i\mathbf{k}E \mp i\boldsymbol{\sigma}\cdot\mathbf{p}_1 + \mathbf{j}m \\ -i\mathbf{k}E \pm i\boldsymbol{\sigma}\cdot\mathbf{p}_1 + \mathbf{j}m \\ -i\mathbf{k}E \mp i\boldsymbol{\sigma}\cdot\mathbf{p}_3 + \mathbf{j}m \end{pmatrix} \begin{pmatrix} i\mathbf{k}E \mp i\boldsymbol{\sigma}\cdot\mathbf{p}_2 + \mathbf{j}m \\ i\mathbf{k}E \pm i\boldsymbol{\sigma}\cdot\mathbf{p}_2 + \mathbf{j}m \\ -i\mathbf{k}E \mp i\boldsymbol{\sigma}\cdot\mathbf{p}_3 + \mathbf{j}m \\ -i\mathbf{k}E \pm i\boldsymbol{\sigma}\cdot\mathbf{p}_2 + \mathbf{j}m \end{pmatrix} \begin{pmatrix} i\mathbf{k}E \pm i\boldsymbol{\sigma}\cdot\mathbf{p}_3 + \mathbf{j}m \\ i\mathbf{k}E \mp i\boldsymbol{\sigma}\cdot\mathbf{p}_3 + \mathbf{j}m \\ -i\mathbf{k}E \pm i\boldsymbol{\sigma}\cdot\mathbf{p}_2 + \mathbf{j}m \\ -i\mathbf{k}E \mp i\boldsymbol{\sigma}\cdot\mathbf{p}_1 + \mathbf{j}m \end{pmatrix}$$

The principle is that the $\mathbf{p}$ term determines which of the three components of a baryon carries an 'active' component of that kind. Let's say $\mathbf{p}_3$ determines 'activity' in the first representation at any given time. Then the first quark has an active electric component, the second an active weak component, and the third active inertial and strong components. This will determine whether these charges are present or absent (1 or 0). How will we decide which component determines 'activity' at any moment? This will be determined by the inertial phase, and this will be governed by the direction of $\boldsymbol{\sigma}$, which is unique to each fermion. Clearly, the labels are arbitrary and can be changed to ensure that gauge invariance is preserved. With the two available signs of $\mathbf{p}$ in the two representations given, all six phases of the strong interaction are simultaneously possible.

The strong charge goes through all possible phases, while the weak and electric charges remain relatively (although not absolutely) fixed on single phases. That is, if we specify that one direction of the $\mathbf{p}$ vector instantaneously contains all the information about the system (here described as 'active'), then we can define this as one of the three axes along which, as we saw in Chapter 5, the three quark momentum components could be aligned. The strong interaction will then take us through all three possible directions, with the 'active' one defined at any moment by coincidence with the one we have described as 'inertial'. However, the weak and electric $\mathbf{p}$ components will only be aligned along that of the 'inertial' (and therefore 'active') one in one of three cases.

In a baryon, the weak and electric phases must be on different quarks. The inertial element again goes through all possible phases in fixing the direction of spin $\boldsymbol{\sigma}$. Mesons have the same structure as baryons, except that they are single fermions combined with the corresponding antifermions and the three phases ('colours') should be considered in a purely temporal (i.e. non-spatial) sequence. The weak and electric charges 'switch on/off' as the phase changes through the components 1, 2, 3.

For free fermions or leptons, the phases are purely the inertial phases. Only the direction of vector properties of $\mathbf{p}$, of course, define a strong phase — the magnitude is determined by the combination of E and m. For free fermions, there is no strong charge because no information is carried about direction, and there is no $SU(3)$ symmetry. Leptons have weak and electric occupancy on the same phase, with a temporal cycle, 1–2–3, as the structure rotates through the three directions involved in $\mathbf{p}$.

$$\begin{matrix} \textit{inertial} \\ \textit{strong} \\ \textit{weak} \\ \textit{electric} \end{matrix} \begin{pmatrix} i\boldsymbol{k}E \pm \boldsymbol{i\sigma}\cdot\mathbf{p}_1 + \boldsymbol{j}m \\ i\boldsymbol{k}E \mp \boldsymbol{i\sigma}\cdot\mathbf{p}_1 + \boldsymbol{j}m \\ -i\boldsymbol{k}E \pm \boldsymbol{i\sigma}\cdot\mathbf{p}_1 + \boldsymbol{j}m \\ -i\boldsymbol{k}E \mp \boldsymbol{i\sigma}\cdot\mathbf{p}_1 + \boldsymbol{j}m \end{pmatrix}$$

We can consider iE, $\boldsymbol{\sigma} \cdot \mathbf{p}$ and m as the respective coefficients for the weak, strong and electric vacuum terms. There are two pseudoscalar terms $\pm iE$; six vector terms $\pm\boldsymbol{\sigma} \cdot \mathbf{p}_1$, $\pm\boldsymbol{\sigma} \cdot \mathbf{p}_2$, $\pm\boldsymbol{\sigma} \cdot \mathbf{p}_3$; and one scalar term m. The weak component switches in such a way as to make iE into $-iE$. The strong switches in such a way as to make $\boldsymbol{\sigma} \cdot \mathbf{p}_1$ convert to $-\boldsymbol{\sigma} \cdot \mathbf{p}_1$, and also to $\boldsymbol{\sigma} \cdot \mathbf{p}_2$; $-\boldsymbol{\sigma} \cdot \mathbf{p}_2$; $\boldsymbol{\sigma} \cdot \mathbf{p}_3$; and $-\boldsymbol{\sigma} \cdot \mathbf{p}_3$. The weak transition involves dipolarity. The strong transition requires a constant rate of change of $\mathbf{p}$, which is equivalent to a linear potential. The electric component preserves m. The respective group structures are $SU(2)$, $SU(3)$ and $U(1)$.

Only iE and $\boldsymbol{\sigma} \cdot \mathbf{p}$ vary, and only $\boldsymbol{\sigma} \cdot \mathbf{p}$ varies in magnitude with phase. The E terms are, therefore, always global. There are two ways of setting up these transitions, both using covariant derivatives for the operators iE and $\boldsymbol{\sigma} \cdot \mathbf{p}$. We can either set up a combination of (pseudo)scalar and vector group generators; or, using the idea that these groups represent the spherical symmetry of a point source, and are covariant, we can replace the scalar and vector parts by scalar potential functions of r, associated with E. In the first case, the scalar parts are scalar phase (Coulomb) terms; the vector parts are the generators that make the individual interactions have the $SU(3)$, $SU(2)$ or $U(1)$ symmetries associated with the $\mathbf{p}$, E and m operators in the nilpotent, or with the direction, handedness or radial magnitude of the angular momentum. In the second case, we can group all terms related to a single particle under a single representation of E as a scalar function of r, which applies globally to the entire state.

The two $SU(2)$ states — filled electric and empty electric background — being global, are automatically set with respect to E. That is, the background is incorporated as the potential producing a scalar or $U(1)$ phase in the E term. In the case of spherical symmetry, this becomes a Coulomb potential. It is possible to combine all the information into a single expression by using the fact that Lorentz invariance, in the case of a purely point source with spherical symmetry, allows us to transfer all the information contained in $\boldsymbol{\sigma} \cdot \mathbf{p}$ to the E term by adding a potential function of r which reproduces the specific aspect of spherical symmetry ($SU(3)$, $SU(2)$ or $U(1)$) incorporated in the covariant part of $\boldsymbol{\sigma} \cdot \mathbf{p}$, that is, the part responsible for the interaction. When the frame is chosen such that all this information is transferred to the E term, then all specific phase information is lost. The rotation of vector $\mathbf{p}$ terms ensures that the strong term is a linear function ($\propto r$).

The scalar nature of m ensures that the electric term is a scalar phase (Coulombic) ($\propto 1/r$). The dipolarity of $\pm iE$ ensures that the weak term is a dipolar equivalent of the scalar phase ($\propto 1/r^3$). These options are also evident as a result of directly applying the condition of spherical symmetry to the fermionic state.

9.2 Phase diagrams

We can picture lepton, baryon and meson charge structures using phase diagrams. In the case of the strong interaction, only one component of angular momentum is well-defined at any moment, and the strong charge appears to act in such a way that the well-defined direction manifests itself by 'privileging' one out of three independent phases making up the complete phase cycle. In a truly gauge invariant system, this can only be accomplished in relative terms. If the weak and electric charges are also related to angular momentum, then the same must apply to them, and the relative 'privileging' of phase can only be defined between the different interactions. We have, here, two options. If the 'privileged' or 'active' phases of E and m (or w and e) coincide with each other, then this also determines the 'privileged' phase of $\mathbf{p}$; the result is no 'privileged' relative phase. Since the strong charge is defined only through the directional variation of $\mathbf{p}$, via a 'privileged' relative phase, a system in which the phases coincide cannot be strongly bound. If, however, they are different, then this information can only be carried through $\mathbf{p}$ (or s), and the strong interaction must be present.

We can imagine the arrangements diagrammatically using a rotating vector to represent the 'privileged' direction states for the charges. Each charge has only one 'active' phase out of three at any one time to fix the angular momentum direction; the symbols e, s, and w here refer to these states, not the actual charges. The vectors may be thought of as rotating over a complete spherical surface. In the case of the quark-based states — baryons and mesons — the total information about the angular momentum state is split between three axes, whereas the lepton states carry all the information on a single axis.

The axes in the diagrams represent both charge states and angular momentum states for leptons, mesons and baryons. As we have seen, each type of charge carries a different aspect of angular momentum (or helicity) conservation; s carries the directional information (linked to $\mathbf{p}$); w carries the sign information (+ or − helicity) (linked to iE); e carries information about magnitude (linked to m). Another way of looking at this is to associate these properties, respectively, with the symmetries of rotation, inversion, and translation, which can also be mapped onto the dihedral symmetries (or rotations around x, y and z axes) of the fundamental parameter group.

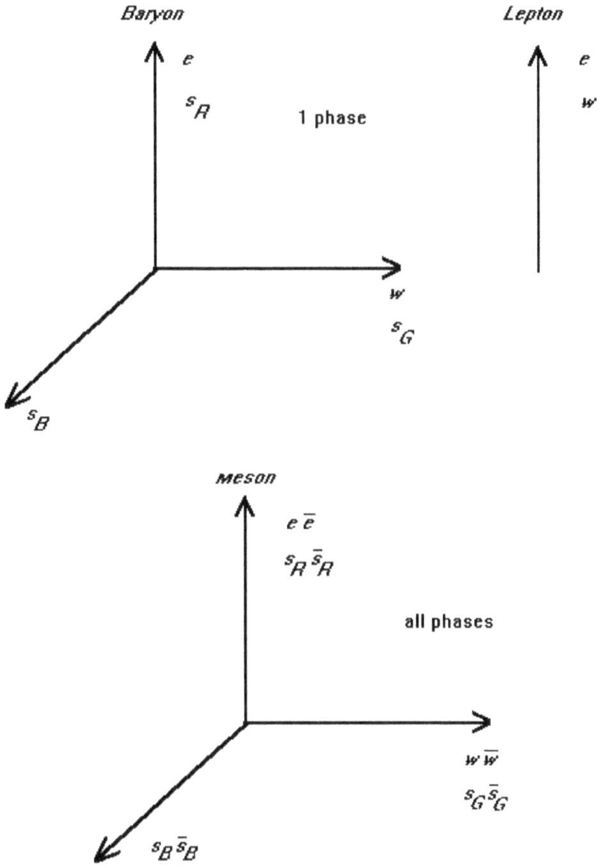

The significance of these diagrams arises from the fact that all the information about the states is contained in the instantaneous direction of the momentum or spin operator $\mathbf{p}$. The absence of a strong interaction for the leptons means that the splitting of this operator into 3 separable parts never takes place, and all the information is along a single direction. In the case of the hadronic or strongly-interacting structures, the information is split between components along 3 orthogonal axes.

9.3 Dirac equation for charge

The $\boldsymbol{k}$, $\boldsymbol{i}$, $\boldsymbol{j}$ operators correspond both to charge (w, s, e) and aspects of angular momentum (E, $\mathbf{p}$, m) (see Chapter 4). So can we use a Dirac-type equation to specifically describe the conservation of charge, rather than of angular momentum? Here we will use one of the more standard forms of the Dirac equation,

$$(\boldsymbol{\alpha} \cdot \mathbf{p} + \beta m - E) \quad \psi = 0,$$

which can be expanded, using a 4×4 matrix, to

$$(\boldsymbol{\alpha} \cdot \mathbf{p} + \beta m - E)\psi = \begin{pmatrix} -E & 0 & im & -ip \\ 0 & -E & ip & im \\ -im & -ip & -E & 0 \\ ip & -im & 0 & -E \end{pmatrix} \begin{pmatrix} \psi_1 \\ \psi_2 \\ \psi_3 \\ \psi_4 \end{pmatrix} = 0.$$

The column vector, here, is the usual 4-component spinor, and the terms E and $\mathbf{p}$ represent the quantum differential operators rather than their eigenvalues.

If we replace the E, $\mathbf{p}$, m terms by the corresponding w, s, e, we can derive an expression for conserved charge:

$$\begin{pmatrix} \boldsymbol{k}w & 0 & -\boldsymbol{ij}e\uparrow & -\boldsymbol{ii}\mathbf{s} \\ 0 & \boldsymbol{k}w & -\boldsymbol{ii}\mathbf{s} & \boldsymbol{ij}e\downarrow \\ -\boldsymbol{ij}e\downarrow & \boldsymbol{ii}\mathbf{s} & -\boldsymbol{k}w & 0 \\ \boldsymbol{ii}\mathbf{s} & \boldsymbol{ij}e\uparrow & 0 & -\boldsymbol{k}w \end{pmatrix} \begin{pmatrix} \boldsymbol{k}w + \boldsymbol{ii}\mathbf{s} + \boldsymbol{ij}e\uparrow \\ \boldsymbol{k}w + \boldsymbol{ii}\mathbf{s} - \boldsymbol{ij}e\downarrow \\ -\boldsymbol{k}w - \boldsymbol{ii}\mathbf{s} + \boldsymbol{ij}e\downarrow \\ -\boldsymbol{k}w - \boldsymbol{ii}\mathbf{s} - \boldsymbol{ij}e\uparrow \end{pmatrix} = 0$$

The 4×4 matrix used here is almost identical in form to the matrix for the Dirac differential operator, although the $+$ and $-$ signs are in different places. The s term effectively takes up the vector-type properties of $\mathbf{p}$, and can be represented as a vector with a single well-defined direction. The electric charge is like mass in the conventional version of the Dirac equation, a passive term. There it is an expression of *zitterbewegung*, here it is weak isospin. The sign applied to e is that of the charge itself, but e has the added property of weak isospin, so that the e's on the first and fourth rows of the matrix and on the first and fourth rows of the column vector can be considered as isospin 'up' ($\uparrow$) and the others as isospin 'down' ($\downarrow$). The opposite states of isospin are not $+$ and $-$ but 1 and 0. So, we should apply to these e terms the matrices:

$$\begin{pmatrix}1\\0\end{pmatrix};\quad \begin{pmatrix}0\\1\end{pmatrix};\quad \begin{pmatrix}-1\\0\end{pmatrix};\quad \begin{pmatrix}0\\-1\end{pmatrix}.$$

The result of this is that all terms involving e disappear on multiplication.

If we now create an exponential term $e^{-i(wt-\mathbf{s}\cdot\mathbf{r})}$, to produce a state vector for charge, and define $i\partial/\partial t = -iw$ and $-i\nabla = i\mathbf{s}$, we obtain:

$$\begin{pmatrix} \boldsymbol{ik}\partial/\partial t & 0 & -\boldsymbol{ij}e\uparrow & -\boldsymbol{i}\nabla \\ 0 & \boldsymbol{ik}\partial/\partial t & -\boldsymbol{i}\nabla & \boldsymbol{ij}e\downarrow \\ -\boldsymbol{ij}e\downarrow & \boldsymbol{i}\nabla & -\boldsymbol{ik}\partial/\partial t & 0 \\ \boldsymbol{i}\nabla & \boldsymbol{ij}e\uparrow & 0 & -\boldsymbol{ik}\partial/\partial t \end{pmatrix} \begin{pmatrix} \boldsymbol{k}w + \boldsymbol{i}i\mathbf{s} + \boldsymbol{ij}e\uparrow \\ \boldsymbol{k}w + \boldsymbol{i}i\mathbf{s} - \boldsymbol{ij}e\downarrow \\ -\boldsymbol{k}w - \boldsymbol{i}i\mathbf{s} + \boldsymbol{ij}e\downarrow \\ -\boldsymbol{k}w - \boldsymbol{i}i\mathbf{s} - \boldsymbol{ij}e\uparrow \end{pmatrix} e^{-i(wt-\mathbf{s}\cdot\mathbf{r})} = 0.$$

The weak isospin terms cancel, suggesting why this becomes the scalar phase term. Without these 'phase' terms, this equation becomes:

$$\begin{pmatrix} \boldsymbol{ik}\partial/\partial t & 0 & 0 & -\boldsymbol{i}\nabla \\ 0 & \boldsymbol{ik}\partial/\partial t & -\boldsymbol{i}\nabla & 0 \\ 0 & \boldsymbol{i}\nabla & -\boldsymbol{ik}\partial/\partial t & 0 \\ \boldsymbol{i}\nabla & 0 & 0 & -\boldsymbol{ik}\partial/\partial t \end{pmatrix} \begin{pmatrix} \boldsymbol{k}w + \boldsymbol{i}i\mathbf{s} \\ \boldsymbol{k}w + \boldsymbol{i}i\mathbf{s} \\ -\boldsymbol{k}w - \boldsymbol{i}i\mathbf{s} \\ -\boldsymbol{k}w - \boldsymbol{i}i\mathbf{s} \end{pmatrix} e^{-i(wt-\mathbf{s}\cdot\mathbf{r})} = 0.$$

Here, each term of the resultant column vector becomes a pseudo-Dirac or Dirac-type equation for charge:

$$(\boldsymbol{ik}\partial/\partial t + \boldsymbol{i}\nabla)(\boldsymbol{k}w + \boldsymbol{i}i\mathbf{s})e^{-i(wt-\mathbf{s}\cdot\mathbf{r})} = 0,$$

in the same way as each term of the resultant column matrix becomes a Dirac equation for the E, $\mathbf{p}$, m combination.

In producing a 'Dirac'-type equation for charge, we are finding *equivalent* properties for the charges to those for the E, $\mathbf{p}$, m terms in a similar way to the one in which equivalent *local* properties reproduced nonlocal ones in Chapter 6. In this interpretation, starting with the real Dirac equation (for E, $\mathbf{p}$, m), we introduce a filled fermion vacuum to create the two-sign degree of freedom required for E. We also define a particular status for antifermions beyond the original requirement that each charge type has two possible signs. We assume, therefore, that a particular type of charge, say s, can only be unit in one of the three 'colours' needed to make up an observed state. This excludes charges of the opposite sign, so we take the concept of antistates from the Dirac equation, and assign $-s$ to the antifermions. We cannot, however, repeat the same procedure for, say, e, which must have both signs in both states and antistates. So, we preserve the rule that a charge ($-e$ in this case) can be unit in only one of the three 'colours', but make the 'default' position (e, e, e) as opposed to (0, 0, 0) for s, and so produce two signs by creating 'weak isospin', with alternatives (e, e, 0) and (0, 0, $-e$). Subsequently, we find that using 'weak isospin' actually gives us a suitable zero for the matrix equation for charge. Finally, to accommodate two signs of w, we have to refer to the fact that a filled vacuum, with antiparticles nonexistent in the ground state, violates charge conjugation symmetry for the charge (w) which specifies the fermionic state.

9.4 Fermionic states from the algebra

Of the total of 64 generators in the Dirac nilpotent algebra, 60 can be arranged into 12 nilpotent pentads, and each of these can be used to represent a fermion. The first term in each pentad represents the energy operator, the next 3 the momentum operator, and the last term the mass operator. If we treat these structures symbolically, they can be seen as representing 12 fermions, say, 6 quarks and 6 leptons, or 6 quarks/leptons and 6 antiquarks/antileptons. The total becomes $2 \times 12 = 24$ if we include left- and right-handed states (the parity P duality); and $2 \times 2 \times 12 = 48$ if we include fermionic and antifermionic states (the charge conjugation C duality), in addition to quarks and leptons.

generation		isospin	
1	electron neutrino	up	$i\mathbf{i}$ $i\mathbf{j}$ $i\mathbf{k}$ ik j
	electron	down	$i\mathbf{i}i$ $i\mathbf{i}j$ $i\mathbf{i}k$ $i\mathbf{k}$ $\mathbf{j}$
2	muon neutrino	up	$j\mathbf{i}$ $j\mathbf{j}$ $j\mathbf{k}$ ii k
	muon	down	$i\mathbf{j}i$ $i\mathbf{j}j$ $i\mathbf{j}k$ $i\mathbf{i}$ $\mathbf{k}$
3	tau neutrino	up	$k\mathbf{i}$ $k\mathbf{j}$ $k\mathbf{k}$ ij i
	tau	down	$i\mathbf{k}i$ $i\mathbf{k}j$ $i\mathbf{k}k$ $i\mathbf{j}$ $\mathbf{i}$

generation		isospin	
1	antielectron-neutrino	up	$-i\mathbf{i}$ $-i\mathbf{j}$ $-i\mathbf{k}$ $-ik$ $-j$
	antielectron	down	$-i\mathbf{i}i$ $-i\mathbf{i}j$ $-i\mathbf{i}k$ $-i\mathbf{k}$ $-\mathbf{j}$

2	antimuon-neutrino	up	−ji −jj −jk −ii −k
	antimuon	down	−iji −ijj −ijk −ii −k
3	antitau-neutrino	up	−ki −kj −kk −ij −i
	antitau	down	−iki −ikj −ikk −ij −i

generation		isospin	
1	up quark	up	ii ij ik ik j
	down quark	down	iii iij iik ik j
2	charm quark	up	ji jj jk ii k
	strange quark	down	iji ijj ijk ii k
3	top quark	up	ki kj kk ij i
	bottom quark	down	iki ikj ikk ij i

generation		isospin	
1	antiup-quark	up	−ii −ij −ik −ik −j
	antidown-quark	down	−iii −iij −iik −ik −j
2	anticharm-quark	up	−ji −jj −jk −ii −k
	antistrange-quark	down	−iji −ijj −ijk −ii −k
3	antitop-quark	up	−ki −kj −kk −ij −i
	antibottom-quark	down	−iki −ikj −ikk −ij −i

9.5 Equation for specifying particle states

It is possible to write down a single equation to generate the entire set of charge structures for quarks and leptons (and their antistates):

$$\boldsymbol{\sigma}_z.\left(\boldsymbol{i}\hat{\mathbf{p}}_a(\delta_{bc}-1)+\boldsymbol{j}(\hat{\mathbf{p}}_b-\mathbf{1}\delta_{0m})+\boldsymbol{k}\hat{\mathbf{p}}_c(-1)^{\delta_{1g}}g\right) \tag{A}$$

The quaternion operators $\boldsymbol{i}$, $\boldsymbol{j}$, $\boldsymbol{k}$ are respectively strong, electric and weak charge units; $\boldsymbol{\sigma}_z$ is the spin pseudovector component defined in the z-direction (here used as a reference); $\hat{\mathbf{p}}_a$, $\hat{\mathbf{p}}_b$, $\hat{\mathbf{p}}_c$ are each units of quantized angular momentum, selected *randomly* and *independently* from the three orthogonal components $\hat{\mathbf{p}}_x$, $\hat{\mathbf{p}}_y$, $\hat{\mathbf{p}}_z$. $\boldsymbol{\sigma}_z$ and the remaining terms are logical operators representing existence conditions, and defining four fundamental divisions in fermionic states. Each of the operators creates one of these fundamental divisions — fermion/antifermion; quark/lepton (colour); weak up isospin/weak down isospin; and the three generations — which are identified, respectively, by the weak, strong, electromagnetic and gravitational interactions.

(1) $\boldsymbol{\sigma}_z = -\mathbf{1}$ defines left-handed states; $\boldsymbol{\sigma}_z = \mathbf{1}$ defines right-handed. For a filled weak vacuum, left-handed states are predominantly fermionic, right-handed states become antifermionic 'holes' in the vacuum (which is 0 in this representation).
(2) $b = c$ produces leptons; $b \neq c$ produces quarks. If $b \neq c$ we are obliged to take into account the three directions of $\mathbf{p}$ at once. If $b = c$, we can define a single direction. Taking into account all three directions at once, we define baryons composed of three quarks (and mesons composed of quark and antiquark), in which each of a, b, c cycle through the directions x, y, z.
(3) m is an electromagnetic mass unit, which selects the state of weak isospin. It becomes 1 when present and 0 when absent. So $m = 1$ is the weak isospin up state; and $m = 0$ weak isospin down. The unit condition can be taken as an empty electromagnetic vacuum; the zero condition a filled one.
(4) g represents a conjugation of weak charge units, with $g = -1$ representing maximal conjugation. If conjugation fails maximally, then $g = 1$. g can also be thought of as a composite term containing a parity element (P) and a time-reversal element (T). So, there are two ways in which the conjugated PT may remain at the unconjugated value (1). $g = -1$ produces the generation u, d, ν_e, e; $g = 1$, with P responsible, produces c, s, ν_μ, μ; $g = 1$, with T responsible, produces t, b, ν_τ, τ.

The weak interaction can only identify (1). This occupies the $\boldsymbol{ik}E$ site in the anticommuting Dirac pentad $(\boldsymbol{ik}E + \boldsymbol{i}\mathbf{p} + \boldsymbol{j}m)$ with the i term being responsible for the fermion/antifermion distinction. Because it is attached to a complex operator, the sign of $\boldsymbol{k}$ has two possible values even when those of $\boldsymbol{i}$ and $\boldsymbol{j}$ are fixed; the sign of the weak charge associated with $\boldsymbol{k}$ can therefore only be determined physically by the sign of $\boldsymbol{\sigma}_z$. The filled weak vacuum is an expression of the fact that the 'ground state of the universe' can be specified in terms of positive, but not negative, energy (E), because, physically, this term represents a continuum state.

The strong interaction identifies (2). This occupies the $\boldsymbol{i}\mathbf{p}$ (or $\boldsymbol{i\sigma}\cdot\mathbf{p}$) site and it is the three-dimensional aspect of the $\mathbf{p}$ (or $\boldsymbol{\sigma}\cdot\mathbf{p}$) term which is responsible for the three-dimensionality of quark 'colour'. A separate 'colour' cannot be identified any more successfully than a separate dimension, and the quarks become part of a system, the three parts of which have $\hat{\mathbf{p}}_a$ values taking on one each of the orthogonal components $\hat{\mathbf{p}}_x$, $\hat{\mathbf{p}}_y$, $\hat{\mathbf{p}}_z$. Mesonic states have corresponding values of $\hat{\mathbf{p}}_a$, $\hat{\mathbf{p}}_b$ and $\hat{\mathbf{p}}_c$ in the fermion and antifermion components, although the logical operators δ_{0m} and $(-1)^{\delta_{1g}}g$ may take up different values for the fermion and the antifermion, and the respective signs of $\boldsymbol{\sigma}_z$ are opposite.

The electromagnetic interaction identifies (3). This occupies the $\boldsymbol{j}m$ site in the Dirac pentad. Respectively, the three interactions ensure that the orientation, direction and magnitude of angular momentum are separately conserved. Gravity (mass), finally, identifies (4).

There is a charge conjugation from $-w$ to w, in the second and third generations, with corresponding violations of P and T symmetries. This is brought about by the filled weak vacuum needed to avoid negative energy states. The two weak isospin states are associated with this idea in (3), the $\mathbf{1}$ in $(\hat{\mathbf{p}}_b - \mathbf{1}\delta_{0m})$ being a 'filled' state, with its absence an unfilled state, and the weak interaction acts by annihilating and creating e, either filling the vacuum or emptying it — which is why, unlike the strong interaction, it always involves the equivalent of particle + antiparticle = particle + antiparticle, and involves a massive intermediate boson. We thus create two possible vacuum states to allow variation of the sign of electric charge by weak isospin, and this variation is linked to the filling of the vacuum which occurs in the weak interaction, and could be connected with a mass-related 'bosonic' spin 0 linking of the two isospin states (in addition to the spin 1 gauge bosons involved in the interaction).

The weak and electric interactions are linked by this filled vacuum in the $SU(2)_L \times U(1)$ model, as they are in the description of weak isospin, and we can regard these as alternative formalisms for representing the same physical truth. It is significant that the Higgs mechanism for generating masses of intermediate weak bosons and fermions requires the same Higgs vacuum field both for $SU(2)_L$ and for $U(1)$. In addition, the combination of scalar and pseudoscalar phases in the mathematical description of the combined electric and weak interactions clearly relates to the use of a complex scalar field in the conventional derivation of the Higgs mechanism.

The formalism actually explains easily how mass is generated when an element of partial right-handedness is introduced into an intrinsically left-handed system. In principle, anything which alters the signs of the terms in the expression $(\boldsymbol{i}\hat{\mathbf{p}}_a(\delta_{bc} - 1) + \boldsymbol{j}(\hat{\mathbf{p}}_b - \mathbf{1}\delta_{0m}) + \boldsymbol{k}\hat{\mathbf{p}}_c(-1)^{\delta_{1g}} g)$, or reduces any of these terms to zero, is a mass generator, because it is equivalent to introducing the opposite sign of $\boldsymbol{\sigma}_z$ or a partially right-handed state. Thus mass can be produced separately by weak isospin, by quark confinement, and by weak charge conjugation violation. The degree of right-handedness is a direct measure of the value of mass, as it determines the *zitterbewegung* frequency. The second and third generations, by successively violating parity and time-reversal symmetry, effectively bring about 'step functions' in this introduction of a right-handed component.

The expression (A) is remarkable in leading to exactly twelve fermionic structures, created by discrete operations with differing degrees of right-handedness introduced.

down	$-\boldsymbol{\sigma} \cdot (-\boldsymbol{j}\hat{\mathrm{p}}_a + \boldsymbol{i}\hat{\mathrm{p}}_b + \boldsymbol{k}\hat{\mathrm{p}}_c)$
up	$-\boldsymbol{\sigma} \cdot (-\boldsymbol{j}(\hat{\mathrm{p}}_a - \mathbf{1}) + \boldsymbol{i}\hat{\mathrm{p}}_b + \boldsymbol{k}\hat{\mathrm{p}}_c)$
strange	$-\boldsymbol{\sigma} \cdot (-\boldsymbol{j}\hat{\mathrm{p}}_a + \boldsymbol{i}\hat{\mathrm{p}}_b - z_P\boldsymbol{k}\hat{\mathrm{p}}_c)$
charm	$-\boldsymbol{\sigma} \cdot (-\boldsymbol{j}(\hat{\mathrm{p}}_a - \mathbf{1}) + \boldsymbol{i}\hat{\mathrm{p}}_b - z_P\boldsymbol{k}\hat{\mathrm{p}}_c)$
bottom	$-\boldsymbol{\sigma} \cdot (-\boldsymbol{j}\hat{\mathrm{p}}_a + \boldsymbol{i}\hat{\mathrm{p}}_b - z_T\boldsymbol{k}\hat{\mathrm{p}}_c)$
top	$-\boldsymbol{\sigma} \cdot (-\boldsymbol{j}(\hat{\mathrm{p}}_a - \mathbf{1}) + \boldsymbol{i}\hat{\mathrm{p}}_b - z_T\boldsymbol{k}\hat{\mathrm{p}}_c)$
electron	$-\boldsymbol{\sigma} \cdot (-\boldsymbol{j}\hat{\mathrm{p}}_a + \boldsymbol{k}\hat{\mathrm{p}}_a)$
e neutrino	$-\boldsymbol{\sigma} \cdot (-\boldsymbol{j}(\hat{\mathrm{p}}_a - \mathbf{1}) + \boldsymbol{k}\hat{\mathrm{p}}_a)$
muon	$-\boldsymbol{\sigma} \cdot (-\hat{\mathrm{p}}\boldsymbol{j}_a - z_P\boldsymbol{k}\hat{\mathrm{p}}_a)$
μ neutrino	$-\boldsymbol{\sigma} \cdot (-\boldsymbol{j}(\hat{\mathrm{p}}_a - \mathbf{1}) - z_P\boldsymbol{k}\hat{\mathrm{p}}_a)$
tau	$-\boldsymbol{\sigma} \cdot (-\boldsymbol{j}\hat{\mathrm{p}}_a - z_T\boldsymbol{k}\hat{\mathrm{p}}_a)$
τ neutrino	$-\boldsymbol{\sigma} \cdot (-\boldsymbol{j}(\hat{\mathrm{p}}_a - \mathbf{1}) - z_T\boldsymbol{k}\hat{\mathrm{p}}_a)$

Both antiquarks and antileptons simply replace $-\boldsymbol{\sigma}$ with $\boldsymbol{\sigma}$.

9.6 Tables of particle structures

The outcome of all the processes determining the 0 and 1 in e, s, w charge structures of the fundamental fermions may be expressed in terms of a set of three 'quark' tables, $A - C$, with an extra table L for the left-handed leptons and antileptons (the unlabelled columns in L represent left-handed antineutrinos). Applying these to the known fermions, $A - C$ would appear to have all the properties of the coloured quark system, with s pictured as being 'exchanged' between the three states (although in reality, of course, all the states exist simultaneously), in the same way as the operator $\mathbf{p}$ in the nilpotent baryon wavefunction. In relation to these tables, we can look on symmetry breaking, in general, as a consequence of the setting up of the algebraic model for charges. When we map time, space and mass onto the charges w, s, e, to create the anticommuting Dirac pentad, only one charge (s) has the full range of vector options. 'Fixing' one of the others (say e) for s to vary against gives us only 2 remaining options for w, unit on the same colour as e or unit on a different one. Putting both w and e on the same colour denies the necessary three degrees of freedom in the direction of angular momentum, so this is forbidden in a quark system. (It is assumed that a violation of parity, symbolised by z_P, or a violation of time-reversal symmetry symbolised by z_T, acts in the second and third generations to restore the sign of the weak charge.)

A

		B	G	R
u	$+e$	$1j$	$1j$	$0i$
	$+s$	$1i$	$0k$	$0j$
	$+w$	$1k$	$0i$	$0k$
d	$-e$	$0j$	$0k$	$1j$
	$+s$	$1i$	$0i$	$0k$
	$+w$	$1k$	$0j$	$0i$
c	$+e$	$1j$	$1j$	$0i$
	$+s$	$1i$	$0k$	$0j$
	$-w$	z_Pk	$0i$	$0k$
s	$-e$	$0j$	$0k$	$1j$
	$+s$	$1i$	$0i$	$0k$
	$-w$	z_Pk	$0j$	$0i$
t	$+e$	$1j$	$1j$	$0i$
	$+s$	$1i$	$0k$	$0j$
	$-w$	z_Tk	$0i$	$0k$
b	$-e$	$0j$	$0k$	$1j$
	$+s$	$1i$	$0i$	$0k$
	$-w$	z_Tk	$0j$	$0i$

B

		B	G	R
u	$+e$	$1j$	$1j$	$0k$
	$+s$	$0i$	$0k$	$1i$
	$+w$	$1k$	$0i$	$0j$
d	$-e$	$0i$	$0k$	$1j$
	$+s$	$0j$	$0i$	$1i$
	$+w$	$1k$	$0j$	$0k$
c	$+e$	$1j$	$1j$	$0k$
	$+s$	$0i$	$0k$	$1i$
	$-w$	z_Pk	$0i$	$0j$
s	$-e$	$0i$	$0k$	$1j$
	$+s$	$0j$	$0i$	$1i$
	$-w$	z_Pk	$0j$	$0k$
t	$+e$	$1j$	$1j$	$0k$
	$+s$	$0i$	$0k$	$1i$
	$-w$	z_Tk	$0i$	$0j$
b	$-e$	$0i$	$0k$	$1j$
	$+s$	$0j$	$0i$	$1i$
	$-w$	z_Tk	$0j$	$0k$

C

		B	G	R
u	$+e$	$1j$	$1j$	$0k$
	$+s$	$0i$	$1i$	$0j$
	$+w$	$1k$	$0k$	$0i$
d	$-e$	$0j$	$0k$	$1j$
	$+s$	$0i$	$1i$	$0k$
	$+w$	$1k$	$0j$	$0i$
c	$+e$	$1j$	$1j$	$0k$
	$+s$	$0i$	$1i$	$0j$
	$-w$	$z_P k$	$0k$	$0i$
s	$-e$	$0j$	$0k$	$1j$
	$+s$	$0i$	$1i$	$0k$
	$-w$	$z_P k$	$0j$	$0i$
t	$+e$	$1j$	$1j$	$0k$
	$+s$	$0i$	$1i$	$0j$
	$-w$	$z_T k$	$0k$	$0i$
b	$-e$	$0j$	$0k$	$1j$
	$+s$	$0i$	$1i$	$0k$
	$-w$	$z_T k$	$0j$	$0i$

L

				ν_e
	$+e$	$1j$	$1j$	$0j$
	$+s$	$0k$	$0i$	$0i$
	$+w$	$0i$	$0k$	$1k$
				e
	$-e$	$0i$	$0k$	$1j$
	$+s$	$0j$	$0i$	$0i$
	$+w$	$0k$	$0j$	$1k$
				ν_μ
	$+e$	$1j$	$1j$	$0j$
	$+s$	$0k$	$0i$	$0i$
	$-w$	$0i$	$0k$	$z_P k$
				μ
	$-e$	$0i$	$0k$	$1j$
	$+s$	$0j$	$0i$	$0i$
	$-w$	$0k$	$0j$	$z_P k$
				ν_τ
	$+e$	$1j$	$1j$	$0j$
	$+s$	$0k$	$0i$	$0i$
	$-w$	$0i$	$0k$	$z_T k$
				τ
	$-e$	$0i$	$0k$	$1j$
	$+s$	$0j$	$0i$	$0i$
	$-w$	$0k$	$0j$	$z_T k$

9.7 $SU(5)$

The idea that the 5-fold Dirac algebra is responsible for the symmetry breaking which leads to the $SU(3) \times SU(2)_L \times U(1)$ splitting in the interactions between fundamental particles suggests that Grand Unification between the three local interactions may involve the $SU(5)$ group, or even $U(5)$, or something containing $SU(5)$ and extending it to right-handed states, such as $SO(10)$. In principle, we effectively have five units expressed in different forms:

E	p_x	p_y	p_z	m
w	s_G	s_R	s_B	e
$\boldsymbol{ik}$	$\boldsymbol{ii}$	$\boldsymbol{ij}$	$\boldsymbol{ik}$	$\boldsymbol{j}$
γ^0	γ^1	γ^2	γ^3	γ^5

The mapping of the strong terms is always exact, but the electroweak terms are so closely linked physically that transposition to equivalent representations may be necessary to reflect the physical manifestations of these interactions. The five charge units (e, s_G, s_B, s_R, w, taking into account the vector nature of s) map directly onto the five Dirac operators ($i\boldsymbol{k}$, $\mathbf{i}\boldsymbol{i}$, $\mathbf{j}\boldsymbol{i}$, $\mathbf{k}\boldsymbol{i}$, $\boldsymbol{j}$), and the five quantities (m, p_x, p_y, p_z, E) involved in the Dirac equation, and generate both an overall $SU(5)$ and its breakdown into $SU(3) \times SU(2)_L \times U(1)$. The 24 $SU(5)$ generators can be represented in terms of any of these units with the bosonic transitions occurring from the values associated with the state with the bar to those associated with the state without. For example:

	$\bar{s}_G$	$\bar{s}_R$	$\bar{s}_B$	$\bar{w}$	$\bar{e}$
s_G					
s_B		gluons		Y	X
s_R					
w		Y		Z^0, γ	W^-
e		X		W^+	Z^0, γ

or

	$\bar{p}_x$	$\bar{p}_y$	$\bar{p}_z$	$\bar{E}$	$\bar{m}$
p_x					
p_y		gluons		Y	X
p_z					
E		Y		Z^0, γ	W^-
m		X		W^+	Z^0, γ

The only unobserved generators here are the strong-electroweak bosons X and Y, which earlier $SU(5)$ schemes have taken to imply direct proton decay. However, such decay would be forbidden by separate charge conservation rules as it involves the complete elimination of a strong charge unit, and it also disregards the necessary dipolarity of the weak charge. Here, the X and Y generators remain linked to the dipolar particle + antiparticle mechanism of the ordinary weak interaction. There is, in fact, one mechanism that unites strong and electroweak interactions in the way that would be expected of X and Y, and this is ordinary beta decay. We can propose that, though this process is not mediated by X or Y at the energies now observable, there will be an amplitude for X or Y intervention at higher energies.

$SU(5)$, however, is not the full story. If we had a 25th generator (which the Standard Model disregards on the grounds that it is not observed), the group would become $U(5)$, and all the generators would be entirely equivalent to scalar phases. Such a particle, if it existed, would couple to all matter in proportion to the amount, and, as a colour singlet, would be ubiquitous. This is a precise description of the gravitational-inertial force of gravity, and, if we can show that Grand Unification of the electromagnetic, strong and weak forces occurs at the Planck mass, the energy usually taken as characteristic of quantum gravity, then it is possible that $U(5)$, or something containing it, might be the true Grand Unification group, and that it also incorporates a gravitational-inertial generator, most probably a spin 1 pseudoboson for the inertial reaction. This generator would then link the 2 colourless gluons with the Z^0 and γ, along the diagonal of the group table, suggesting a link between all four interactions. Reduction of the generators to scalar phases would mean that, at Grand Unification, all interactions would be identical in effect, and all non-Coulombic structure would disappear. The unification would be exact.

9.8 Quarks

The non-Abelian gauge theory of quantum chromodynamics (QCD) for the strong interaction between coloured quarks has turned out to be one of the most successful aspects of the Standard Model. In the model as normally applied, the up quark has electric charge $2e/3$, where e is the fundamental electric charge, while the down quark has a charge of $-e/3$. The 'strong' property manifested in the baryon number (B), equivalent to our strong charge, is shared out between the 'coloured' quarks.

	Blue	**Green**	**Red**
up	$2e/3$	$2e/3$	$2e/3$
	$B/3$	$B/3$	$B/3$
down	$-e/3$	$-e/3$	$-e/3$
	$B/3$	$B/3$	$B/3$

The phenomenology of quantum electrodynamics has shown over many experiments that quarks behave as though they are constituted in exactly this way, irrespective of the energy of the interaction. The question is, why does nature choose a structure like this? We seem to have 3 separate 'units' of charge ($e/3$, $2e/3$ and e), with an implication that the electron, with e, and even the up quark (with $2e/3$), may not actually be 'elementary'. What we seem to see here is evidence of a *broken symmetry*. Now, broken symmetries, according to our fundamental thinking, can't exist at the most fundamental level. They are *emergent*. So what could be more fundamental? To my way of thinking, it has to be something that puts quarks and leptons on the same level and shows their common origins.

In fact, there is such a possibility almost as old as Gell-Mann and Zweig's original quark theory of 1963, and that is the first coloured quark theory by Han and Nambu of 1964, which proposed that exactly the same results could be obtained using *integral* and zero charges and assigning an integral baryon number to a single quark.[27] Now this proposal appears to fit particularly well with the structures which seem to emerge from the nilpotent wavefunction, with the baryon number taking the place of the momentum operator. Here, colour is not an additional property, but an integral part of the structure, and the quarks now look significantly like leptons.

	Blue	**Green**	**Red**
up	e	e	0
	B	0	0
down	0	0	$-e$
	B	0	0

Broken symmetries seem to emerge where the situation introduces complexity. Here, the complexity for the electric charge distribution is the strong interaction which is the only reason for the quarks' existence, and which demands perfect gauge invariance. This cannot be broken and the quarks in either representation will necessarily be seen as fractional for QED, so showing the correctness of the first version *for that purpose*. The difference between the two representations can now be clarified. In the first, the fractional nature of electric charges is due to the electric interaction itself; in the second it is externally imposed by the strong interaction. Because the masslessness of gluons shows that the gauge invariance of this interaction is exact, there will never be a high energy regime where the integral nature of the charges will be revealed. The fractional electric charges reflect the perfect equivalence between the different coloured states or phases of the interaction. They are QED or electroweak eigenstates.

We now know how the process works in the parallel case of the fractional quantum Hall effect, where electrons appear with effective charges of $e/3$, $e/5$ and other fractional values, because an electron or other fermion can form a pseudobosonic combination with an odd number of magnetic flux lines and so effectively share itself out between them.[28] Interestingly, in this parallel case, it is the *weak* interaction, rather than the strong, acting as the external agent determining QED phenomenology. The question is not whether QED should use fractional charges — clearly it should — but whether these are *fundamental* or stem from a broken symmetry. QED phenomenology clearly doesn't decide the question of which basic structure of charges to use in other areas such as Grand Unification, or the *gauge relations between the interactions.* Using our fundamental methodology would suggest that we go for the best fit with the symmetry, and, if that leads to modifications in our usual practice, see if this has any interesting consequences. In fact, as we shall see, there is one of exceptional interest.

9.9 Grand Unification: a prediction

A Grand Unified Theory (GUT), to unite electric, weak and strong interactions, based on the $SU(5)$ group, was first proposed by Georgi and Glashow in 1974, following on from the Glashow–Weinberg–Salam $SU(2) \times U(1)$ unification of the electric and weak interactions.[29] The electroweak unification is governed by the weak mixing angle parameter $\sin^2\theta_W$, which is effectively the ratio (α/α_2) between the weak and electric couplings (α_2 and α). Georgi and Glashow showed that, in any GU scheme determined by a single GU gauge group, $\sin^2\theta_W$ would be given by the ratio of the sum of all the squared units of weak isospin (t_3) for the fermions of the Standard Model to the sum of all their squared units of electric charge (Q).

$$\sin^2\theta_W = \frac{Tr(t_3^2)}{Tr(Q^2)}.$$

Taking the weak components with only left-handed contributions to weak isospin, for the first generation of quarks and leptons, that is, for 3 colours of u, 3 colours of d, and the leptons e and ν, we obtain:

$$Tr(t_3^2) = \frac{1}{4} \times 8 = 2.$$

Quarks and leptons have identical units of weak isospin, and so this summation will be the same for both quark theories, and will also be the result expected for phenomenology. But for the electric charge structure, the summations of the two representations is different. For fractional charges with both left- and right-handed contributions in the first generation, we obtain

$$Tr(Q^2) = 2 \times \left(\frac{4}{9} \times 3 + \frac{1}{9} \times 3 + 1 + 0 \right) = \frac{16}{3}$$

from which

$$\sin^2 \theta_W = 0.375.$$

For integral charges, however, we have

$$Tr(Q^2) = 2 \times (1 + 1 + 0 + 0 + 0 + 1 + 1 + 0) = 8,$$

leading to

$$\sin^2 \theta_W = 0.25.$$

The value of 0.375 is, as Steven Weinberg has said, in 'gross disagreement' with the experimental value for $\sin^2\theta_W$ of 0.231 at around the mass-energy of the Z particle ($M_Z = 91\,\text{GeV}$),[30] but 0.25 is relatively close to this value and would be even closer (with some small second order corrections) if the effect of the direct production of W and Z bosons at their mass-energies M_W and M_Z is taken into account (or if the 0.25 occurs at the vacuum expectation energy (246 GeV) rather than at M_W or M_Z). 0.25 is also the value that would be obtained purely from the leptonic contribution, and it is rather curious that the value for a purely electroweak parameter should be different in the quark and lepton sectors.

Going back to the original 'minimal $SU(5)$' GUT of Georgi and Glashow, we find that it doesn't actually unify the pure interactions at all, for, though the theory begins with the equations for the running weak and strong coupling constants, derived by quantum field theory from their respective $SU(2)$ and $SU(3)$ structures:

$$\frac{1}{\alpha_2(\mu)} = \frac{1}{\alpha_G} - \frac{5}{6\pi} \ln \frac{M_X^2}{\mu^2},$$

$$\frac{1}{\alpha_3(\mu)} = \frac{1}{\alpha_G} - \frac{7}{4\pi} \ln \frac{M_X^2}{\mu^2},$$

(where M_X is the GU energy scale, α_G is the fine structure constant at this energy and μ is the energy scale of measurement) it proposes that the Grand Unified gauge group structure will modify the equivalent $U(1)$

equation for the electromagnetic coupling ($1/\alpha$), assumed (in this theory) to be

$$\frac{1}{\alpha(\mu)} = \frac{1}{\alpha_G} + \frac{5}{3\pi} \ln \frac{M_X^2}{\mu^2},$$

to one in which it is mixed with the weak value, based on $SU(2) \times U(1)$, where

$$\frac{5}{3\alpha_1(\mu)} = \frac{1 - \sin^2 \theta_w}{\alpha}.$$

From these equations, we derive a Grand Unified mass scale (M_X) of order 10^{15} GeV, and from

$$\sin^2 \theta_W = \frac{\alpha(\mu)}{\alpha_2(\mu)},$$

we find 'renormalized' values of $\sin^2\theta_W$ at the measurement scale of order 0.19 to 0.21.

As is well known, the curves representing the variations of the parameters α_1, α_2 and α_3 at different energy scales (μ) don't actually cross at anything very close to a point, leading to the somewhat *ad hoc* proposal that a supersymmetric model may be the only solution. In addition we are forced to use a combined electroweak parameter which makes assumptions about group structure, and relies on a particular value for the squared 'Clebsch–Gordan coefficient' of the group, $C^2 = 1/\sin^2 \theta_W - 1 = 5/3$, that has, as yet, no experimental or theoretical justification. Unifying *electroweak*, weak and strong parameters doesn't seem as convincing as using the original electric, weak and strong values, while the assumed value of $\sin^2\theta_W = 0.375$ at GU suggests that the electroweak unification is not even then complete, as the two forces are not on equal footing.

The 'convergence' is also three or four orders of magnitude below the scale of the Planck energy at which quantum gravity is assumed to operate, suggesting that another principle will be needed to include gravitation. But, in any case, compensating errors in the combination tend to disguise the massive inconsistencies between the separate equations for the coupling constants. In particular, recalculation of the value of $\sin^2\theta_W$ at $\mu = 10^{15}$ GeV gives 0.6 rather than the 0.375 which was initially assumed in setting up the equations!

Now, one advantage of using integral quark charges is that it gives us an *independent* value for $\sin^2\theta_W = \alpha/\alpha_2$ of the right order, and we can perform a much simpler calculation for M_X without making assumptions about the group structure, by avoiding the problematic running coupling constant equation for $1/\alpha_1$, using only the more secure equations for $1/\alpha_2$ and $1/\alpha_3$. In addition, the hypercharge numbers for the $U(1)$ electromagnetic running coupling equation will now be no longer identical to those for a quark model based purely on QED phenomenology. The fermionic contribution to QED vacuum polarization is, for *fundamental* fractional charges,

$$\frac{4}{3}\times\frac{1}{2}\times\left(\frac{1}{36}\times 3+\frac{1}{36}\times 3+\frac{1}{9}\times 3+\frac{4}{9}\times 3+\frac{1}{4}\times 1+\frac{1}{4}\times 1+1\right)\frac{n_g}{4\pi}=\frac{5}{3\pi},$$

where $n_g = 3$ is the number of fermion generations, and the terms in the bracket represent, respectively, the squared average charge in the isospin quark doublet, the squared charges of the quarks, the squared average charge of the isospin lepton doublet, and the squared charges of the leptons, all for both left- and right-handed states. Modifying this for fundamental integral charges, we obtain:

$$\frac{4}{3}\times\frac{1}{2}\times\left(\frac{1}{4}\times 3+\frac{1}{4}\times 3+1+1+0+0+0+1+\frac{1}{4}\times 1+\frac{1}{4}\times 1+1\right)\frac{n_g}{4\pi}$$
$$=\frac{3}{\pi}.$$

This result corresponds to a change in the squared Clebsch–Gordan coefficient from $C^2 = 5/3$ to $C^2 = 3$, when $\sin^2\theta_W = 1/(1 + C^2)$ changes from 0.375 to 0.25. With the new values we have obtained for the hypercharge numbers, the running coupling of the pure electromagnetic interaction will be:

$$\frac{1}{\alpha(\mu)}=\frac{1}{\alpha_G}+\frac{3}{\pi}\ln\frac{M_X^2}{\mu^2}.$$

Leaving out the speculative equation for $1/\alpha_1$, and, for the moment, this new one for $1/\alpha$, but using the well-established ones for $1/\alpha_2$ and $1/\alpha_3$, and $\sin^2\theta_W = \alpha/\alpha_2$, we obtain

$$\sin^2\theta_W(\mu)=\alpha(\mu)\left(\frac{1}{\alpha_3(\mu)}+\frac{11}{6\pi}\ln\frac{M_X}{\mu}\right).$$

Taking typical values for $\mu = M_Z = 91.2\,\text{GeV}$, $\alpha_3(M_Z^2) = 0.118$ (or 0.12), $\alpha(M_Z^2) = 1/128$, and $\sin^2\theta_W = 0.25$, we obtain a value for the GU energy scale ($M_X = 2.8 \times 10^{19}\,\text{GeV}$) which is extraordinarily close to the Planck value ($1.22 \times 10^{19}\,\text{GeV}$), and may well be exactly so, as purely first-order calculations overestimate the value of M_X. Assuming that M_X *is* the Planck mass, we obtain α_G (the GU value for all interactions) $= 1/52.4$, and

$\alpha_2(M_Z^2) = 1/31.5$, which is exactly the kind of value we would expect for the weak coupling with $\sin^2\theta_W = 0.25$ close to M_Z.

To provide an independent check on the validity of the procedure, we can *now* make direct use of the equation we have derived for $1/\alpha$, with the new hypercharge numbers and GU at the Planck mass, to obtain $1/\alpha(M_Z^2) = 128$, which is, of course, exactly the value obtained experimentally at energies corresponding to $\mu = M_Z$. This appears to be a striking confirmation of the assumptions made in the first calculation leading to M_X, as coincidental agreements are most unlikely for equations involving logarithmic terms, and it is also potentially very significant, for it would now appear that the unification which occurs at M_X might well involve a direct numerical equalization of the strengths of the three, or even four, physical force manifestations, without reference to the exact unification structure.

The analysis suggests that, at Grand Unification, $C^2 = 0$ and $\sin^2\theta_W = 1$, creating an exact symmetry in every respect between weak and electric interactions, as well as between weak and strong, which is completely different from the only partial unification achieved using the fundamental fractional charges, and linking this with the scale associated with quantum gravity. The mixing parameter, $\sin^2\theta_W$, as normally understood, may then be interpretable as the electroweak constant for a specifically *broken* symmetry, taking the value of 0.25 at the energy range where the symmetry breaking occurs (presumably at $M_W - M_Z$, or, alternatively, the expectation value of the Higgs field, 246 GeV), and gradually decreasing from the maximum ($\sin^2\theta_W = 1$) to this value at intermediate energies. At GU, we may suppose, all four forces are reduced to scalar phases, with $U(1)$ symmetry and purely Coulombic interaction, all distinguishing aspects of the weak and strong interactions having diminished to zero.

One of the most significant aspects of the calculation is that it leads to completely testable predictions, as the values of the three coupling constants can be calculated for any energy with relative precision from the known values of α_G and M_X. In particular, the value of α changes rapidly in a way that can be determined at energies now available to us experimentally. At 14 TeV, for instance, it would have the value of $1/118$, compared to $1/125$ from the minimal $SU(5)$ theory of Georgi and Glashow.

Relatively simple considerations based on results from the extensive quantum field theories of the electric, weak and strong interactions, which require only a small amount of arithmetical and algebraic manipulation, thus suggest that, if the fundamental representation behind the fractional broken symmetry requires integral charges, then it has major consequences for GU, which are accessible by experiment, and it would also have another advantage in making quark–lepton unification (in symmetry terms) much more likely, as both sectors would now be characterised by integral charges, and quarks would have the same electroweak structure as leptons.

9.10 The Higgs mechanism and fermion masses

There is another area where this application of fundamental methodology may have a significant consequence. This is in the problematic application of the Higgs mechanism to the generation of fermion masses. Here, to generate separate masses for the two isospin states in each generation, we require two different hypercharge (or $2 \times$ average charge) units of 1 and -1, yet, if the fractional charges are fundamental, there is only one hypercharge value for all quarks, and that is the fractional value, $2/3$. The only expedient then is to 'invent' two hypercharges not justified by the assumed charge structure. If integral charges are fundamental, however, the different colours of quark automatically produce the two hypercharge values, 1 and -1, which we require for both isospin states and which would be repeated in each generation.

The leptons, of course, are not fractionally charged, but there is a separate area of difficulty with them. In the past, the lepton mass mechanism could be accommodated by assigning the single hypercharge value in the first generation to electrons, but the discovery of neutrino masses means that the opposite hypercharge value is now required for neutrinos. It is possible that this difficulty can be resolved in both representations if the neutrino is a Majorana particle with a low mass resulting from the low probability of the neutrino transforming to its antistate with the opposite hypercharge.

9.11 Larger group structures for fermions and bosons

One of the major questions in particle physics is: can the fundamental particles be linked in a single group representation? The particles represent many broken symmetries, and, according to our methodology, broken symmetries are a result of complexity or synthesis, and not of some unknown 'symmetry-breaking principle' which applies to large-scale structures. There appear to be two major symmetries in the parameter group based on the numbers 2 (for duality) and 3 (for anticommutativity), and versions of these link up in the creation of fermion point particles in a broken symmetry based on the number 5.

If we look at the fundamental particles, all the symmetries which apply to them seem to be constructed from smaller symmetries based on these units. The same also applies to many of the groups thought to be of significance in this area, particularly those based on the octonion symmetries, such as the exceptional groups E_6, E_7, and E_8. Because the symmetry breaking is ultimately 3-dimensional in origin (and manifested, for example, in quarks and 3 particle generations), the symmetries involved in particle groupings tend to map naturally onto geometries in 3-dimensional space. However, the higher groupings which collect together particles such as quarks and leptons, or fermions and bosons, also show relationships with structures in higher-dimensional spaces and groups connected with them. Here, as we would expect, the symmetries become less broken, perhaps culminating in an unbroken root vector structure in E_8, the highest group symmetry to emerge from this type of mathematics. Generally, the structures in the higher-dimensional figures carry with them the numbers associated with those from the lower dimensions.

The group E_8, has long been suspected of being a possible unifying group for the fundamental particles, and was discussed as such, among other places, in *Zero to Infinity*. In 2007, Garrett Lisi proposed that all known fermions and gauge bosons could be fitted into the 240 root vectors of the E_8 group.[31] The model has been heavily criticized, and doesn't look right as it stands. Its particles don't add up to 240, leading to a completely *ad hoc* speculation about particles needed to make up the numbers, the gravity theory is very speculative, the generations don't arise naturally, *etc.* Some of the assignments seem very difficult to understand. Though I don't think that the model is correct, the idea may be, and it fits in with previous ideas on the significance of E_8. One of the things that was criticized was the inclusion of fermions and bosons in the same representation. Lisi argued that this was possible, but only through the exceptional groups E_6, E_7, E_8. If this is correct (and it has gained some support), it may be Lisi's most significant contribution, along with the emphasis on root vectors.

In the Standard Model, the particles are divided into fermions (spin $\frac{1}{2}$) and gauge bosons (spin 1), and the fermions are divided into quarks and leptons. There are 6 quarks arranged in 3 generations, each of which has 2 weak isospin states (up/down; charm/strange; top/bottom), and each of which comes in 3 varieties of 'colour'. Corresponding to these are 6 leptons, again in 3 generations, each with 2 weak isospin states (electron neutrino/electron; muon neutrino/muon; tau neutrino/tau). There are no colours associated with the leptons, so each set of 3 coloured quarks and 1 lepton, in each isospin state in each generation, represents a kind of 4-dimensional structure, parallel to that of space and time. The total of real fermions in the Standard Model is therefore 24 (18 coloured quarks + 6 leptons).

In addition, each fermion has 2 possible spin states, and to every fermionic state there is a corresponding antifermion state, making a total of 96 real fermionic + antifermionic states. The 2 spin states and fermion/antifermion options are an intrinsic aspect of the fermion's spinor structure, $(\pm i\boldsymbol{k}E \pm \boldsymbol{i}\mathbf{p} + \boldsymbol{j}m)$. Now, spin 1 gauge bosons can be represented by $(\pm i\boldsymbol{k}E \pm \boldsymbol{i}\mathbf{p} + \boldsymbol{j}m)\ (\mp i\boldsymbol{k}E \pm \boldsymbol{i}\mathbf{p} + \boldsymbol{j}m)$. In effect, 4 fermionic states are required to produce a boson, and the nilpotent formalism shows that 1 spin 1 vacuum boson (never seen, but still mathematically necessary) is created for every 4 real fermionic states, and 4 vacuum fermionic states (again never seen, but still necessary) are created for every real spin 1 boson.

Now, the number of real spin 1 gauge bosons in the Standard Model is 12 (8 gluons, W^+, W^-, Z^0, γ), but virtually all Grand Unified theories (and certainly $SU(5)$) predict the existence of another 12 ($6\,X$ and $6\,Y$) to unify strong and electroweak interactions. These are equivalent in number but are distinct from the 24 vacuum bosons which accompany the 96 real fermionicic/antifermionic states in the nilpotent formalism, in the same way as the real fermionic/antifermionic states are distinct from the 96 vacuum fermionic/antifermionic states which accompany the 24 real bosons. A combination of 96 fermions/antifermions and 24 spin 1 gauge bosons creates a total of 120 real particle states. If there are an equal number of vacuum states, then the total becomes 240, the kissing number in 8 dimensions and also the number of root vectors in E_8. Perhaps we can justify something like the following:[32]

	Quarks	Leptons	Bosons		Fermions	Bosons			Extra options compared to 1 generation with 1 isospin and 1 spin				
1	3	1	1	=	4	1	=	5					
2	6	2	2	=	8	2	=	10	S				
3	9	3	3	=	12	3	=	15					G
4	12	4	4	=	16	4	=	20	S	I			
5	18	6	6	=	24	6	=	30	S				G
6	24	8	8	=	32	8	=	40	S	I	A		
7	36	12	12	=	48	12	=	60	S	I			G
8	48	16	16	=	64	16	=	80	S	I	A	V	
9	72	24	24	=	96	24	=	120	S	I	A		G
10	144	48	48	=	192	48	=	240	S	I	A	V	G

The successive rows in the table represent the increased options provided by spin, isospin, antistates, vacuum and three generations:

1	1 generation with 1 isospin and 1 spin
2	1 generation with 1 isospin and 2 spins
3	3 generations with 1 isospin and 1 spin
4	1 generation with 2 isospins and 2 spins
5	3 generations with 1 isospin and 2 spins
6	1 generation with 2 isospins and 2 spins + antiparticles
7	3 generations with 2 isospins and 2 spins
8	1 generation with 2 isospins and 2 spins + antiparticles + vacuum
9	3 generations with 2 isospins and 2 spins + antiparticles
10	3 generations with 2 isospins and 2 spins + antiparticles + vacuum

Notably absent from this structure are the spin 0 Higgs boson (which is not a gauge boson), the spin 2 graviton (which may not exist), and the spin 1 inertial pseudoboson (which is not really a separate particle from the photon, but a special realisation of it at the Planck energy).

Particles constitute a 5 of 3 quarks + lepton + boson, which multiplies by 4 factor 2 dualities (spin up/down, isospin up/down, fermion/antifermion, particle/vacuum, symbolised respectively by S, I, A, V, and relating, respectively, to space, charge, time and mass) and 1 factor 3 triplet (generations, symbolised by G), and so (depending on the order in which the 4 factors of 2 and 1 of 3 are multiplied) generates number series up to a maximum number of 240. All products of 5 are equally artificial constructs, for example, linking fermions with bosons in the exceptional groups E_6 to E_8 through the fact that the last term in the 5 can be a scalar.

In effect, we invert the derivation of 12 structures from a 5-unit pentad, and map the fermions and bosons onto a new pentad structure, of which the pseudoscalar component (the iE term) is 24 leptons/antileptons, and the vector component (the $\mathbf{p}$ term) 72 quarks/antiquarks. Bosons are scalar particles, and scalars are the squared products of pseudoscalars and vectors, just as bosons are the squared products of fermions/antifermions. So, if the 24 bosons occupy the *scalar* part of the pentad (the m term), then we can use nilpotency to group the 96 fermions (24 leptons and 72 quarks) with the 24 bosons into a single structure with 120 fermions plus bosons, and these would seem to be represented by the stages $48 + 12 = 60$, $96 + 24 = 120$, $192 + 48 = 240$. So, in a nilpotent system, we have a physical as well as mathematical reason for combining fermions and bosons in the same representation, and, in the nilpotent structure, a fermion is necessarily, in some sense, its own vacuum boson, and vice versa. The 12 bosons we would put with the 48 are the 8 gluons, the 2 W and 1 Z boson, and the photon, that is, all except X and Y.

The interesting thing about the whole idea is that we can follow through many of the fundamental algebras derived from real and complex numbers, quaternions and octonions, geometries in spaces from 3 to 8 dimensions, and many associated groups, and the key numbers all seem to appear in this table. In it, we see the complexity building up from the simplest symmetries in a way that suggest why the higher symmetries have physical meaning and why they are always broken.

Chapter 10

Return to Symmetries

10.1 A universal rewrite system

The seemingly unbroken consistency of the symmetries described in Chapters 3 and 4 gives a strong indication that they are close to a true foundational level in physics, and the apparent applicability of foundational methods in resolving problems in relatively complex parts of physics gives support to this. If this is a valid inference, there is still one major question to be answered: does foundations of physics have a foundation? And, if so, what kind? Clearly, it would also have to act in some way as a foundation for mathematics, and possibly for a more general approach to information processing.

The twentieth century success of digital computing made a number of people think that nature might be structured on an information system, similar to the workings of a computer. So, can this be true here? Modern computers are based on the Turing machine, digital logic, and the rewrite or production system, which is essentially the basis of programming and algorithmic processes of any kind. So, we have a pretty clear idea of how they work. A conventional rewrite system has 4 fixed components:

alphabet
rewrite rules (productions)
a start 'axiom' or symbol
stopping criteria

As an example (suggested by my computer scientist colleague, Bernard Diaz) we can devise a process for generating Fibonacci numbers. There are two rules: p1: A $\rightarrow$ B; p2: B $\rightarrow$ AB. We begin with generation 0, and a single symbol A. Rule p1 then tells us to replace A with B. In the next generation,

B becomes AB by rule p2. From generation 2, the A in AB becomes B, while the B becomes AB, so we end up with BAB in generation 3, and the process then repeats indefinitely:

		length of string
N=0	A	1
N=1	→B	1
N=2	→AB	2
N=3	→BAB	3
N=4	→ABBAB	5
N=5	→BABABBAB	8
N=6	→ABBABBABABBAB	13
N=7	→BABABBABABBABBABABBAB	21

It may be significant that the rules A → B and B → AB seem to be suggesting the structure of 3-dimensional (quaternion) algebra:

$$\boldsymbol{i} \rightarrow \boldsymbol{j} \quad \boldsymbol{j} \rightarrow \boldsymbol{ij}(\text{“} = \boldsymbol{k}\text{”})$$

and that a string like BABABBABABBABBABABBAB appears to be creating a fractal-like structure in 3-dimensional space, but situated in the AB or $\boldsymbol{ij}$ plane, as in holography. The logarithmic spiral then becomes a way of expressing 3-dimensionality in the plane with the increasing length of the intervals substituting for penetration into the third dimension.

Now, if we had a system in which all four elements — alphabet, start object, rules, and stopping criteria — could be varied, the rewrite structure and alphabet would become universal, and perhaps self-generating. The algebraic structure which we believe lies at the foundations of physics has the appearance of being generated by a process, possibly an information process. Can we use the concept of totality zero to find the process or the algorithm? On the basis of the foundational principles we have declared in the first chapter, should we expect it to be universal?

No universal rewrite system being known, Bernard Diaz and I set out to see if nature actually supplied one, and if it looked like what was driving physics and mathematics as set out in the Klein-4 parameter group and the nilpotent package and algebraic structure used for fermionic physics. In this search it is important not to assume any prior knowledge of mathematics, not even the natural numbers. This is very difficult to do, as counting has become a basis for virtually everything we do, and, even if we avoid actual counting, is difficult to avoid the use of it in our language.

This is what we came up with. We assume that the universe, or any alphabet which it contains, is always a zero totality state, with no unique description, and so infinitely degenerate. This means that we have to continually regenerate the alphabet, in such a way that it is always new, but there is no limit or stopping criterion, as the new state created is always another nonunique zero totality. It is a kind of *zero attractor*. A nonzero deviation from 0 (say R) will always incorporate an automatic mechanism for recovering the zero (say the 'conjugate' R^*), but the zero totality which results, say (R, R^*), will not be unique, and will necessarily lead to a new structure.

The characteristics of the process which emerge from this include self-similarity, scale-independence, duality, bifurcation, and holism. The process differs from other rewrite processes in having no fixed starting or ending point, and an alphabet and production rules that are endlessly reconstructed during the process. So, it fulfils our conditions for universality. The self-similarity is a necessary consequence of the lack of a fixed starting point. It suggests that, if there are physical applications at one level, then there are likely to be applications also at others. As we scale up from small to larger systems, we can imagine that some principle such as the renormalization group takes effect to maintain the form of the structures generated by the rewrite process.

We ensure that a structure or alphabet is new by defining the position of all previous structures within it as subalphabets. The process will then continue indefinitely. Effectively, the process involves defining a series of *cardinalities*. Successive alphabets absorb the previous ones in the sequence, so creating a new cardinality. The cardinalities are like Cantor's cardinalities of infinity, but are cardinalities of zero instead. From the point of view of the observer, i.e. someone 'inside' the system ('universe', 'nature', 'reality'), we have to start from (R, R^*), which is the minimum description of a zero totality alphabet (or a zero totality universe in physical terms). The successive stages are all zeros as we go from one zero cardinality or totality to the next, and we ensure that they are cardinalities by always including the previous cardinality or alphabet. So (R, R^*, A, A^*), for example, includes (R, R^*). We may start at any arbitrary zero-totality alphabet but there is no natural beginning or end to the process. Because all the stages are cardinalities or zero totality alphabets, the process is always holistic. We have to include everything.

A convenient though not unique way of representing the process is by a 'concatenation' or placing together, with no algebraic significance, of any given alphabet with respect to either its components or subalphabets or itself. Since an alphabet is defined to be a cardinality, then anything other than itself must necessarily be a 'subalphabet' and the concatenation will produce nothing new. Only concatenation of the entire alphabet with itself will produce a new cardinality or zero totality alphabet. It is convenient to represent these two aspects of the process by the symbols $\Rightarrow$ for *create*, in which every alphabet produces a new one which incorporates itself as a component, and $\rightarrow$ for *conserve*, which means that nothing new is created by concatenating with a subalphabet.

conserve: (subalphabet) (alphabet) $\rightarrow$ (alphabet) *there is nothing new*

create: (alphabet) (alphabet) $\Rightarrow$ (new alphabet) *a zero totality is not unique*

The process is simultaneously recursive, creating everything E (all symbols) at once, and iterative, creating a single symbol only. It is also fractal and can begin or end at any stage. Since we intend that it should describe or create both time and 3D space, we can think of it as prior to both. The create and conserve aspects must also be simultaneous; we only know which new alphabet will emerge when we have ensured that all possible concatenations with subalphabets yield only the alphabet itself.

Suppose, we then start with a zero totality alphabet of the form (R, R^*). Of course, we have to assume that this is not necessarily the beginning, though it is the point where we as observers start from. So, this already 'bifurcated' state will have started from a previous alphabet, which we assume we can't access directly, because we have no structure for it. If we describe this as R, then the * or R^* character creates the 'doubling' process. (Although we now see this in terms of the number 2, we have to imagine that we have not yet invented this method of labelling, and the same applies to the term 'duality'.) Before we create (R, R^*), we have to assume that (R) is a zero totality alphabet, but it is a zero to which we have no access. In effect, we are trying to posit an ontology that exists before the epistemology or observation, begins with (R, R^*). So, we assume that it must happen without being able to observe it.

Applying the conserve process ($\rightarrow$) to concatenate (R, R^*) with its subalphabets should produce nothing new. No concept of 'ordering' is needed in this process, but each term must be distinct. So

$$(R)(R, R^*) \rightarrow (R, R^*)$$

$$(R^*)(R, R^*) \rightarrow (R^*, R) \rightarrow (R, R^*)$$

It follows immediately that these concatenations lead to rules of the form:

$$(R)(R) \to (R); (R^*)(R) \to (R^*);$$
$$(R)(R^*) \to (R^*); (R^*)(R^*) \to (R)$$

The next stage is to show that the zero-totality alphabet $(R,\ R^*)$ is not unique, and that a concatenation with itself will produce a new zero-totality alphabet. Our first suggestion might be something like $(A,\ A^*)$, but, with the terms undefined, this is indistinguishable from $(R,\ R^*)$, and so the only way to ensure that the new alphabet is distinguishable from the old is by incorporating the old one, and we need to do this in such a way that we ensure that the subalphabets yield nothing new. So we try

$$(R, R^*)(R, R^*) \Rightarrow (R, R^*, A, A^*) \tag{1}$$

Having used the 'create' mechanism, we now apply the conserve operation $(\to)$ to this new alphabet, and concatenate with the subalphabets. So

$$(R)(R, R^*, A, A^*) \to (R, R^*, A, A^*) \to (R, R^*, A, A^*)$$
$$(R^*)(R, R^*, A, A^*) \to (R^*, R, A^*, A) \to (R, R^*, A, A^*)$$
$$(A)(R, R^*, A, A^*) \to (A, A^*, R^*, R) \to (R, R^*, A, A^*)$$
$$(A^*)(R, R^*, A, A^*) \to (A^*, A, R, R^*) \to (R, R^*, A, A^*)$$

As before the order of the terms is different for each operation, as we require, but the total is the same, and we soon quickly realise that (R, R^*) and (A, A^*) can only be different if

$$A\,A \to R^*, \text{ etc.,} \quad \text{while } R\,R \to R.$$

Duality is intrinsic to the process. The operation $(\,)(\,) \Rightarrow (\,,\,)$ describes how we go from one zero totality alphabet — or description of the universe — to the next one up. The $(\,,\,)$ is a kind of 'doubling' or 'bifurcation'. So we could write the result of (1) in the form $(R,\ R^*,\ A,\ A^*)$, and represent $(R,\ R^*)$ $(R,\ R^*)$ as a kind of doubling, to create a new cardinality $(R,\ R^*,\ A,\ A^*)$, almost like transforming the second $(R,\ R^*)$ into $(A,\ A^*)$.

The next stage presents a new problem, for

$$(R, R^*, A, A^*)(R, R^*, A, A^*) \Rightarrow (R, R^*, A, A^*, B, B^*)$$

would fail the application of the conservation mechanism $(\to)$ by introducing new concatenated *terms* like AB, AB^*, which lie outside the alphabet. This means that we must include these in advance, as in

$$(R, R^*, A, A^*)(R, R^*, A, A^*) \Rightarrow (R, R^*, A, A^*, B, B^*, AB, AB^*).$$

After this, the question that remains is: does this new alphabet satisfy all our requirements, when we concatenate separately with (R), (R^*), (A), (A^*), (B), (B^*), (AB), (AB^*)? The process is straightforward for the first six concatenations:

$$(R)(R, R^*, A, A^*, B, B^*, AB, AB^*) \rightarrow (R, R^*, A, A^*, B, B^*, AB, AB^*)$$
$$(R^*)(R, R^*, A, A^*, B, B^*, AB, AB^*) \rightarrow (R^*, R, A^*, A, B^*, B, AB^*, AB)$$
$$(A)(R, R^*, A, A^*, B, B^*, AB, AB^*) \rightarrow (A, A^*, R^*, R, AB, AB^*, B, B^*)$$
$$(A^*)(R, R^*, A, A^*, B, B^*, AB, AB^*) \rightarrow (A^*, A, R, R^*, AB^*, AB, B^*, B)$$
$$(B)(R, R^*, A, A^*, B, B^*, AB, AB^*) \rightarrow (B, B^*, AB, AB^*, R^*, R, A, A^*)$$
$$(B^*)(R, R^*, A, A^*, B, B^*, AB, AB^*) \rightarrow (B^*, B, AB^*, AB, R, R^*, A^*, A)$$

But concatenations of the *concatenated terms*, (AB) and (AB^*), on themselves and on each other appear to leave us with two options, which we can describe as 'commutative' and 'anticommutative':

$$(AB)(AB) \rightarrow (R) \ (\textit{commutative})$$
$$(AB)(AB) \rightarrow (R^*) \ (\textit{anticommutative})$$

In fact, however, there is no choice, for *only the anticommutative option* produces something new. Labelling is arbitrary in the rewrite structure, and so the labels A and B alone cannot distinguish these terms from each other — this can only be done if they produce distinguishable outcomes. The commutative option leaves A and B indistinguishable except by labelling, and so does not extend the alphabet. We are obliged to default on the anticommutative option, which means that the last two concatenations become:

$$(AB)(R, R^*, A, A^*, B, B^*, AB, AB^*) \rightarrow (AB, AB^*, B, B^*, A, A^*, R^*, R)$$
$$(AB^*)(R, R^*, A, A^*, B, B^*, AB, AB^*) \rightarrow (AB^*, AB, B^*, B, A^*, A, R, R^*)$$

This solution for A and B cannot be repeated to include new terms, such as (C), $(D), \ldots$, when we extend the alphabet to higher stages because an inconsistency will always reveal itself at some point in the analysis. Anticommutativity produces a closed 'cycle' with components (A, B, AB) and their conjugates (A^*, B^*, AB^*), and prevents further terms, such as C, D, *etc.*, from anticommuting with them in a consistent manner. However, successive cycles of the form (A, B, AB), (C, D, CD), *etc.*, can be introduced into the structure, if they commute with each other, and this can be continued indefinitely. All of the terms then have a unique identity *because they* *each have a unique partner*, and the successive alphabets can be seen as a regular series of identically structured closed anticommutative cycles, each of which commutes with all the others.

This structure is familiar to us in the form of the infinite series of finite (binary) integers of conventional mathematics, each alphabet representing a new integer. We can regard the closed cycles as an infinite ordinal sequence, and so establishing for the first time in this process both the number 1 and the binary symbol 1 of classical Boolean logic as a conjugation state of 0, with the alphabets structuring themselves as an infinite series of binary digits. Mathematics and digital logic become emergent properties of a rewrite process which has no specific defined starting point, and can be reconstructed endlessly in a fractal manner with self-similarity at all stages and a zero attractor.

The universal rewrite system is a pure description of process with many representations. It can, for example, be presented as a series of 'doublings' or 'bifurcations', analogous to the initial creation of (R, R^*), even when they represent 'complexification' (the introduction of a new anticommutative cycle) or 'dimensionalization' (the closing of the cycle) rather than conjugation (the zeroing process used in the first alphabet). So, one way of writing

$$(R, R^*, A, A^*)(R, R^*, A, A^*) \Rightarrow (R, R^*, A, A^*, B, B^*, AB, AB^*)$$

would be as

$$(R, R^*, A, A^*)(R, R^*, A, A^*) \Rightarrow (R, R^*, A, A^*, \quad B, B^*, AB, AB^*)$$

in which we retain the old alphabet (R, R^*, A, A^*) and a dual (B, B^*, AB, AB^*) formed by some process, as we did with R and R^*. In practical terms, we introduce a new character (B).

10.2 Mathematical representation

A significant representation is the mathematical one, for the effective creation of a discrete integer system means that, by applying this to the original terms (R, A, B, *etc.*), we can also generate an entire arithmetic and an algebra. Once we have generated integers, the rest of the constructible number system will follow automatically along with arithmetical operations. At the same time, application of the constructed number systems to the undefined state with which the process began suggests that this state, which is not intrinsically discrete, can be interpreted in terms of a continuity of real numbers in the Cantorian sense.[33]

In principle, discreteness appears in the construction only when we introduce anticommutativity or 'dimensionality', and specifically 3-dimensionality. Physically, 3-dimensionality, or anticommutativity, becomes the ultimate source of discreteness in a zero totality universe, and we observe, in all cases, that 3-dimensionality requires discreteness, and discreteness requires 3-dimensionality.

The rewrite process can be represented in symbolic form in the table:

	0	Δ_a	Δ_b	Δ_c	...	Δ_n
0	00	$0\Delta_a$	$0\Delta_b$	$0\Delta_c$		$0\Delta_n$
Δ_a	$\Delta_a 0$	$\Delta_a\Delta_a$	$\Delta_a\Delta_b$	$\Delta_a\Delta_c$		$\Delta_a\Delta_n$
Δ_b	$\Delta_b 0$	$\Delta_b\Delta_a$	$\Delta_b\Delta_b$	$\Delta_b\Delta_c$		$\Delta_b\Delta_n$
Δ_c	$\Delta_c 0$	$\Delta_c\Delta_a$	$\Delta_c\Delta_b$	$\Delta_c\Delta_c$		$\Delta_c\Delta_n$
:						
Δ_n	$\Delta_n 0$	$\Delta_n\Delta_a$	$\Delta_n\Delta_b$	$\Delta_n\Delta_c$		$\Delta_n\Delta_n$

Here, the Δ symbols represent the alphabets:

Δ_a (R, R^*)
Δ_b (R, R^*, A, A^*)
Δ_c $(R, R^*, A, A^*, B, B^*, AB, AB^*)$
Δ_d $(R, R^*, A, A^*, B, B^*, AB, AB^*, C, C^*, AC, AC^*, BC, BC^*, ABC, ABC^*)$
...

The process is not restricted to any specific mathematical interpretation, but incorporates digital logic and binary integers, while a convenient consequence is the algebraic series:

$$(1, -1)$$
$$(1, -1) \times (1, i_1)$$
$$(1, -1) \times (1, i_1) \times (1, j_1)$$
$$(1, -1) \times (1, i_1) \times (1, j_1) \times (1, i_2)$$
$$(1, -1) \times (1, i_1) \times (1, j_1) \times (1, i_2) \times (1, j_2)$$
$$(1, -1) \times (1, i_1) \times (1, j_1) \times (1, i_2) \times (1, j_2) \times (1, \boldsymbol{i}_3) \ldots$$

The anticommutative pairs A, B; C, D; E...now become successive quaternion units, i_1, j_1; i_2, j_2; $\boldsymbol{i}_3$ ..., each of which is commutative to all the others. By the fourth stage, we have repetition, which then continues indefinitely. An incomplete set of quaternion units (for example, $\boldsymbol{i}_3$ in the sixth alphabet) becomes equivalent to the algebra of complex numbers. Mathematically, we can see the process of the creation of the zero totality alphabets as one of conjugation, followed by repeated cycles of complexification and dimensionalization.

At the point where the cycle repeats, we have what can be recognised as a Clifford algebra — the algebra of 3D space, where the vectors i, j, k are constructed from $i_2\boldsymbol{i}_3$, $j_2\boldsymbol{i}_3$, $i_2 j_2\ \boldsymbol{i}_3$, and i_1, j_1, $i_1\ j_1 = k_1$ and i_2, j_2, $i_2\ j_2 = k_2$ are (mutually commutative) quaternion algebras of the form i, j, k.

$$
\begin{aligned}
&(1, -1)\\
&(1, -1) \times (1, i)\\
&(1, -1) \times (1, i) \times (1, j)\\
&(1, -1) \times (1, i) \times (1, j) \times (1, \mathrm{i})\\
&(1, -1) \times (1, i) \times (1, j) \times (1, \mathrm{i}) \times (1, \mathrm{j})\\
&(1, -1) \times (1, i) \times (1, j) \times (1, \mathrm{i}) \times (1, \mathrm{j}) \times (1, \boldsymbol{i}) \ldots
\end{aligned}
$$

Standard Clifford vector algebra notably produces these subalgebras in the reverse order to the universal rewrite system, which generates, in its first four alphabets, scalars, pseudoscalars, quaternions and vectors, along with the scalar subalgebras of pseudoscalars and quaternions.

Significantly, if we take all these algebras as *independently true*, and hence commutative, as the rewrite structure seems to suggest we should, since each is a complete description of zero totality, then we require an algebra that is a commutative combination of vectors, bivectors, trivectors and scalars, or vectors, quaternions, pseudoscalars and scalars. This turns out to be equivalent to the algebra of the sixth alphabet, a group structure of order 64 with elements:

$\pm\mathrm{i}$	$\pm\mathrm{j}$	$\pm\mathrm{k}$	$\pm i\mathrm{i}$	$\pm i\mathrm{j}$	$\pm i\mathrm{k}$	$\pm i$	± 1
$\pm i$	$\pm j$	$\pm k$	$\pm ii$	$\pm ij$	$\pm ik$		
$\pm\mathrm{i}i$	$\pm\mathrm{i}j$	$\pm\mathrm{i}k$	$\pm i\mathrm{i}i$	$\pm i\mathrm{i}j$	$\pm i\mathrm{i}k$		
$\pm\mathrm{j}i$	$\pm\mathrm{j}j$	$\pm\mathrm{j}k$	$\pm i\mathrm{j}i$	$\pm i\mathrm{j}j$	$\pm i\mathrm{j}k$		
$\pm\mathrm{k}i$	$\pm\mathrm{k}j$	$\pm\mathrm{k}k$	$\pm i\mathrm{k}i$	$\pm i\mathrm{k}j$	$\pm i\mathrm{k}k$		

These, as we have seen, generate an algebra which is isomorphic to that of the gamma matrices of the Dirac equation, as used in relativistic quantum mechanics. Mathematically, it represents the commutative combination of the first complete Clifford algebra (that of 3D space) with all its subalgebras. So, in many respects, the sixth alphabet represents a particularly significant stage in the rewrite process, the first at which the repetitive nature of the sequence becomes fully established, and this is, therefore, in effect the rewrite system order code.

10.3 Physical application

In physical terms, the first four alphabets suggest the successive emergence of descriptions of the universe in terms of the fundamental parameters mass, time, charge and space, which are described by the algebras created with these alphabets.

$(1, -1)$	real	mass
$(1, -1) \times (1, \boldsymbol{i}_1)$	complex	time
$(1, -1) \times (1, i_1) \times (1, j_1)$	quaternion	charge
$(1, -1) \times (1, i_1) \times (1, j_1) \times (1, \boldsymbol{i}_2)$	vector	space

The simultaneous existence of these independent descriptions of the universe can only be accomplished at the level of the sixth alphabet. This is also the first level at which the processes of conjugation, complexification and dimensionalization, which are successively introduced with the first three alphabets, can be finally incorporated into a repetitive sequence. It is like commutativity and anticommutativity, where the first continues generating possibilities to infinity, while the second closes after a limited finite sequence of three terms.

Since such terms as R, A, B, C, *etc.*, are generated independently of each other by a unique process, no relative numerical 'values' are automatically attached to them, and it is possible, when we use one of the mathematical representations, to choose values for the terms of this alphabet in such a way that its self-product becomes zero and all subsequent alphabets are automatically zeroed as well. This is the nilpotent condition. It can be achieved in an infinite number of ways, each of which is unique, and it appears to apply both to quantum physics and to systems at much higher levels. The nilpotent condition defines its conjugate as a kind of mirror reflection of itself rather than an intrinsically new state, or defined new symbol — physically, we call it 'vacuum'. Significantly, the 'conjugate' state to the creation of the initial symbol 1 in binary arithmetic is not defined by a new symbol either, but by a kind of number 'vacuum'. In binary, $1 = 1$, but $-1 = \ldots 1111111111111111111$.

Now, the rewrite process is a fractal one in which alphabets at one level can become the units at another. We can therefore imagine an interpretation of the rewrite structure in which every new symbol, say A, B, $C, \ldots$ represents the creation of a new nilpotent alphabet. Since the units are now already defined as unique, the connections between them require only the generation of commutative algebraic coefficients, which can then continue to infinity. This suggests the creation of something like a Hilbert space of nilpotent alphabets, defined by a commutative Grassmann algebra. The coefficients increase with every new nilpotent alphabet, but they have no intrinsic significance, serving only to distinguish one nilpotent alphabet from another.

10.4 Entropy and information

This section, codified in a paper written with Peter Marcer,[39] presents entropy as a concept that occurs as a result of the universal rewrite structure independently of any specific application to physics. Because every alphabet always includes the previous one, the universal rewrite system is intrinsically irreversible. Every new alphabet is then always necessarily an extension of the last in the form of a bifurcation. This gives us the simplest possible definition of entropy. A bifurcation at every new creation means that the system will generate 2^n components at the nth stage. Now, the standard definition of entropy, $S = \int dQ/T = k \ln W$, where W is the number of available microstates, only includes k because a measurement of temperature was defined, in historical terms, on the basis of the macroscopic properties of water. However, temperature T never appears in physics without k, and has no meaning separate from the definition of the energy term kT. The true measure of standard entropy, as a fundamental concept independent of any specific method of measurement would be $\int dQ/kT = \ln W$, which is just a number.

With this established, we can simply redefine entropy, say S', as the number n, the order of the alphabet from any given beginning. Then, the classical measure of entropy S becomes $kn \ln 2$, only differing from S' by a choice of numerical factor: $S' = S/k \ln 2$. At the nth stage from the arbitrary beginning, the number of equally probable microstates will be 2^n which we can take as a measure of the increasing complexification/disorder resulting from the process. This will be true for any application of the rewrite process at any level — quantum physics, classical physics, chemistry, biology, the brain, social organization — and to macrosystems of every kind. Our experience that entropy always increases can be taken as evidence that the rewrite system is always bifurcating The rate of entropy increase then becomes a measure of the bifurcation that actually occurs, as well as a standard clock, while observation of the entropy increase becomes an indication of the way in which the rewrite process operates in any given system, as well as allowing us to establish that it is a universal process.

Digital logic and information in the computational sense may be seen as a specialised result of the universal rewrite procedure, and so this definition of entropy would be in agreement with Shannon's view of 1948 that at least $kT \ln 2$ must be dissipated per transmitted bit of information, if communication in linear systems is done by waves with part of the energy of which the message consists being dissipated, and so lost from the message. A paper by Rolf Landauer on 'Energy requirement in communication' (1987) begins with this then-accepted view.[34] A minimum dissipation would also be an information measure. Self-organization enables us to consider the loss as a measure, not simply of the loss, but also of the information which allows the self-organization to proceed. The universal rewrite construction shows that the self-organization can be treated as a complete binary tree of ordinal measure with a natural canonical power series structure. At each scale, the measure of information transfer and entropy increase is determined from the level of the alphabet reached in the universal rewrite process.

Now, the global parameter temperature appears in the radiation formula as kT because it concerns the summation over each linearly independent wave phenomenon with phase θ across the entire universe, or from zero to infinity, which is how k, $\hbar$ and c become fundamental global scaling constants of the universal rewrite system. The information in nilpotent quantum mechanics is not, in fact, lost as a result of the constant bifurcation, but rather accumulated to make the global process irreversible, passing from the nonlocal or indistinguishable to the local or distinguishable. Time asymmetry occurs in translations from nonlocal to local effects because the local requires asymmetric, time-like and consequently irreversible solutions, whereas the nonlocal does not. As we have seen, all nonlocal processes also have local manifestations, and this explains how the time component of the vacuum manifests itself in local effects. If an event is observable, it is necessarily in the present, for the future remains as part of the unobservable or nonlocal vacuum.

Referring to the bifurcation process which is the manifestation of the universal rewrite system, we can say that the rate at which it happens must be proportional to the (free) energy involved. The higher the energy, the higher the rate of bifurcation events. Near chaotic systems, involving nonlinearity and high connectivity of the components, transfer energy at near maximum efficiency, and so bifurcate rapidly, generating a correspondingly large measure of entropy. This is especially true of biological systems, which have evolved to be highly organized and composed of many interconnecting parts. Rapid information transfer and states of high entropy become strongly correlated. The process in general acts as a 'clock', with the time interval determined by the rate at which the available options are doubled. The fact that all natural systems are entropic, and irreversible in time, is evidence that all act in terms of the universal rewrite process.

The growth of a chaotic system, like an event in quantum mechanics, provides a perfect parallel with the universal rewrite system. It has been observed that in a typical situation leading to chaos, say the growth of an animal population, there comes a point at which, when the growth rate increases above a certain value, the equation produces a bifurcation between two possible outcomes. Further increases in the growth rate produce a series of further bifurcations of each bifurcation at a frequency determined by a single scale factor for preserving self-similarity, the universal Feigenbaum number 4.669 .... This can be seen as a characteristic extension of an alphabet by the creation of a new one, exactly as in the 'create' process involved in universal rewrite.

The rewrite system describes the evolution of a *process* rather than a physically-defined system, though the process might itself require a bifurcation in the system. In effect, a near-chaotic system becomes subject to particularly rapid overall change because of its high degree of nonlinearity and interconnectivity, and the bifurcation occurs at the level of the whole system or the process applied to the whole system, rather than in only a part of it. When the process applies collectively, rather than to just part of the system, the expansion of the rewrite structure leads to a complete bifurcation or doubling of the options, and we would expect this to happen repeatedly. A system operating near chaos with a high degree of nonlinearity and connectedness of its parts will have a high efficiency in transferring free energy, and be subject to rapid development. The existence of a universal scale factor in the outcome may be taken as a consequence of the relative holism of a system or process on the edge of chaos.

The process also relates to the growth of complexity in natural systems. In principle, nature creates systems and objects whose most required state is self-annihilation with the rest of the universe, or the universal vacuum.

Everything in nature constantly strives towards this end, resulting in local combinations of systems with the nearest available manifestation of a process tending towards this, e.g. fermions and antifermions. Since complete zeroing isn't possible locally, the result is complexity and combinations where the symmetry is imperfect or broken, and where the parts continually strive to make further connections. The nearest to 'stability' occurs when an object combines with an 'environment' that fulfils a maximal approximation to the desired vacuum connection (e.g. a nucleus in an atom, a bulk molecule in condensed matter, a cell in an organism, an aerobic bacterium being absorbed within an archaeum). To create and maintain this pseudovacuum environment requires a maximal number of connections and interactions to be made and maintained, and hence a maximal generation of entropy. Since the most 'desired' state is the combination (and consequent annihilation) of any system with the universal background, then the tendency of the evolution of the universe will always be in the direction of maximum entropy.

10.5 Duality and the factor 2

Both physics and mathematics encompass a fundamental principle of duality at their very bases. Essentially, this is how we create 'something from nothing'. If the ultimate thing that we wish to describe is really 'nothing', then we can only create 'something' as part of a dual pair, in which each thing is opposed by another thing which negates it. We can describe this mathematically in terms of the simplest known symmetry group (C_2), which is essentially equivalent to an object and its mirror image (or 'dual'), whose components are the positive and negative versions of a quantity which may be left undefined.

This has a surprisingly simple manifestation everywhere as the factor 2 or $\frac{1}{2}$, which sometimes becomes equivalent to squaring or square-rooting. Of course, duality does not always imply equal status, and may incorporate *chirality*, as in the different status of $+$ and $-$ units in binary numbering. Duality, in addition, is not a single operation, and the process requires indefinite extension, in the form $C_2 \times C_2 \times C_2 \times \ldots$. If we begin with a unit, there will be an infinite series of 'duals' to this unit via a process which must be carried out with respect to all previous duals (that is, that the entire set of characters generated becomes the new 'unit') and the total result must be zero at every stage.

There are essentially 3 dualities in the parameter group:

nonconserved/conserved, imaginary/real, commutative/anticommutative

Physical dualities always emerge from one or more of these, and they are often interchangeable.

Examples of the first include action + reaction, absorption + emission, radiation + reaction, potential v. kinetic energy, relativistic v. rest mass, uniform v. uniformly accelerated motion, and even rectangles v. triangles. It manifests itself in the use of pairs of *conjugate variables* to define a system, in both classical and quantum physics.

Examples of the second include bosons v. fermions, electric and magnetic fields in Maxwell's equations, and space–like v. time–like systems. It allows transformations to be made, for example, between space and time representations. It is the one which occurs in relativistic contexts. A more subtle form of it occurs in the creation of massive particle states at the expense of components of charge.

Examples of the third include fermion + 'environment' (Aharonov–Bohm, Berry phase, Jahn–Teller, *etc.*), space-like v. time-like systems, particles v. waves, Heisenberg v. Schrödinger/the harmonic oscillator, quantum mechanics v. stochastic electrodynamics/zero-point energy; 4π v. 2π rotation, and all cases in which physical dimensionality or noncommutativity is involved.

So the factor 2 may be seen, for example, as a result of action and reaction (A); commutation relations (C); absorption and emission (E); object and environment (O); relativity (R); the virial relation (V); or continuity and discontinuity (X). The colour coding comes from the fundamental duality from which it emerges. Many of these explanations overlap in the case of individual phenomena, suggesting that they are really all part of some more general overall process:

Kinematics						V	X
Gases	A					V	
Orbits	A					V	X
Radiation pressure	A		E			V	
Gravitational light deflection					R	V	
Fermion/boson spin		C		O	R	V	
Zero-point energy	A	C				V	X
Radiation reaction	A		E		R	V	
SR paradoxes	A		E				

The factor 2 seems to work mainly in one direction. So, the constant terms produce effects which are $2\times$ the changing terms, the real produce ones which are $2\times$ the imaginary, and the discrete produces ones which are $2\times$ the continuous: the multiplication occurs in the direction which doubles the options. The first combines + and – cases where it remains constant; the second involves squaring imaginary parameters to produce real ones; and the third combines dimensionality and noncommutativity with discreteness, and so doubles the elements. However, doubling of options in one direction may be balanced by halving the options in another. The factor appears when we look at a process from a one-sided point of view, and the complete description of a system tends to lead to the overall elimination of the factor. The use of the factor 2 is a two-way process, and the system can only be described in complete terms by taking both the halving and doubling into account. Physical phenomena involving the factor tend to incorporate, in some form, the opposing sets of characteristics.

It will be interesting to look at some of the many cases, especially where there is obvious crossover. One is the Argand diagram, where 2 dimensions of space become a different duality between real and imaginary. In pure geometry, the area of a triangle, $\frac{1}{2}\times$ length of base $\times$ perpendicular height, translates into the graphical representation of kinematics as motion under uniform acceleration (a) as a straight-line $v-t$ graph. Area under the graph is distance travelled, $\frac{1}{2}\,vt$. Under *uniform* v, the distance would have been given by the area of a rectangle, vt. The factor 2 distinguishes steady conditions and steadily *changing* conditions. Under uniform acceleration, we have the 'mean speed theorem', $s=\frac{1}{2}(u+v)t$ and $v^2=u^2+2as$. A more general variation gives us $\frac{1}{2}\,mv^2$ for kinetic energy and $p^2/2m$. Even more generally, we have the virial theorem, $V=2T$. Kinetic energy gives the action side of Newton's third law, potential energy concerns both action and reaction. Newton derived mv^2/r for centripetal force, or mv^2 for orbital potential energy, by having the satellite object being 'reflected' off the circle of the orbit in a polygon with an increasing number of sides, which, in the limiting case, becomes a circle. The imagined physical reflection, by doubling the momentum through action and reaction, then produces the potential, rather than kinetic, energy formula.

A real reflection of ideal gas molecules off the walls of a container produces a momentum doubling, which indicates steady-state conditions, though it is immediately removed by the fact that we have to calculate the average time between collisions ($t=2a/v$) as the time taken to travel *twice* the length of the container (a). The average force then becomes the momentum change/time = 2 $mv/t = mv^2/a$, and the pressure due to one molecule in a cubical container of side a becomes mv^2/a^3, or mv^2/V (volume), leading, for n molecules, to the direct pressure–density relationship, which we call Boyle's law ($P = \rho\bar{c}^2/3$, where $\bar{c}$ is the root mean square velocity). The kinetic behaviour of the ideal gas molecules is actually irrelevant to the derivation since the system describes a steady-state dynamics with positions of molecules constant on a time-average. Taking into account the three dimensions between which the velocity is distributed, the ratio of pressure and density (P/ρ) is derived from the *potential energy* term mv^2 for each molecule and is equal to one third of the average of the squared velocity, or $\bar{c}^2/3$.

Photons, which, unlike material particles, are relativistic objects, surprisingly behave in exactly the same way in a 'photon gas', producing a radiation pressure of the form $P = \rho c^2/3$, with the relativistic energy $E = mc^2$ behaving exactly like a classical potential energy term, and with no mysterious 'relativistic factor' at work. We can consider the photons as being reflected off the walls of the container in exactly the same way as the molecules of materials although the real process obviously also involves absorption and re-emission. In addition, even though free photons have no kinematics, it is also perfectly possible to treat photons acting under the constraint of certain forces as though they have. This is why it is possible to use the standard Newtonian escape velocity equation to derive the Schwarzschild limit for a black hole, with no transition to a 'relativistic' value.

As we have seen in Chapter 9, we can derive the full double gravitational bending of light using the kinetic equation for orbit creation, rather than the potential energy equation used for steady-state conditions. Of course, we *can* use both special and general relativity to derive the effect, but the cause of the effect is independent of the particular version of physics we use to calculate it. In every case where a 'relativistic' correction (either special or general) seems to 'cause' the doubling of a physical effect, the relativistic aspect, like classical kinetic energy, is providing a way of incorporating the effect of *changing conditions* if we begin with the potential, rather than the kinetic, energy equation. Authors have had conflicting views about the doubling, but if we use the potential energy equation where the kinetic energy equation is appropriate, or *vice versa*, then we can find correct physical reasons for almost *any* additional term which doubles the effect predicted.

The same applies to the anomalous magnetic moment or, equivalently, the gyromagnetic ratio of a Bohr electron acquiring energy in a magnetic field, and so ultimately spin $\frac{1}{2}$. Here, using an equation for steady-state conditions, we find only half the measured value, but a relativistic effect (the Thomas precession) doubles the value. But, if we use a kinetic energy equation for changing conditions, for example, at the instant we 'switch on' the field, then we get the correct value immediately. The equation, $\frac{1}{2}m(\omega^2 - \omega_0^2) = e\omega_0 B$, in fact, is only a disguised version of the kinematic equation $v^2 - u^2 = 2as$. Spin is neither relativistic nor quantum in origin, though it can be derived using relativistic or quantum theories — we have already seen that spin $\frac{1}{2}$ can be derived from the anticommutativity of $\mathbf{p}$ as much as from the addition of the Thomas precession. When it is derived from the Schrödinger equation, it is simultaneously derived from the classical kinetic energy term, and, at the same time, produced by the anticommuting nature of the momentum operator.

Applying the Schrödinger equation to the quantum harmonic oscillator requires a *varying* potential energy term $\frac{1}{2}\, m\omega^2 x^2$, taken directly from the classical kinetic energy term $\frac{1}{2}\, mv^2$. The $\frac{1}{2}$ in this expression leads by direct derivation to the $\frac{1}{2}$ in the expression for the ground state or 'zero-point' energy of the system. The same zero-point energy relates to $\hbar/2$ in the Heisenberg uncertainty principle, though the factor $\frac{1}{2}$ there is also generated by anticommutativity in the same way as for electron spin.

The fact that the factor 2 in spin states, which establishes a distinction between bosons and fermions, can be shown to originate ultimately in the virial relation between kinetic and potential energies, has fundamental significance with respect to the fermion/vacuum duality. Kinetic energy is always associated with rest mass m_0 undergoing a continuous change, potential energy is associated with 'relativistic' mass because this term is actually *defined* through a potential energy-type expression ($E = mc^2$), and this implies an equilibrium state with an 'environment', requiring a discrete transition for any change.

The particle and its 'environment' are two 'halves' of a more complete whole. For a material particle, when we expand its relativistic mass-energy term (mc^2) to find its kinetic energy ($\frac{1}{2}\, m_0 v^2$), we either take the relativistic energy conservation equation as a 'relativistic' mass or potential energy equation, incorporating the particle and its interaction with its environment, and then quantize to a Klein–Gordon equation with integral spin; or, we separate out the kinetic energy term using the rest mass m_0 by taking the square root of $E^2 = \gamma^2 m_0^2 c^4$ to obtain $E = m_0 c^2 + m_0 v^2/2 + \cdots$, and, if we choose, quantize to the Schrödinger equation and spin $\frac{1}{2}$.

The $\frac{1}{2}$ occurs in the act of square-rooting, or the splitting of 0 into two nilpotents in the Dirac equation; the $\frac{1}{2}$ in the nonrelativistic Schrödinger approximation is a manifestation of this which we can trace through the $\frac{1}{2}$ in the relativistic binomial approximation. If we go directly to the Dirac equation to obtain the spin $\frac{1}{2}$ term, we see that the same result emerges from the behaviour of the anticommuting terms; the anticommuting property is a direct result of taking the quaternion state vector as a nilpotent. So the anticommuting and binomial factors have precisely the same origin.

The connection between spin and statistics becomes obvious: square-rooting the scalar (and, so, commutative) operator, associated with an integral spin state, to produce two spin $\frac{1}{2}$ states requires the introduction of quaternion operators which are necessarily anticommutative. So particles with integral spins (bosons) follow the Bose–Einstein statistics associated with commutative (symmetric) wavefunctions, while particles with $\frac{1}{2}$-integral spins (fermions) follow the Fermi–Dirac statistics associated with anticommutative (antisymmetric) ones.

The factor 2 also links the continuous with the discontinuous. Expressions involving half units of $\hbar$, representing an average or integrated increase from 0 to $\hbar$, are characteristic of continuous aspects of physics, while those involving integral ones are characteristic of discontinuous aspects. The Schrödinger theory is an example of a continuous option, while the Heisenberg theory is discontinuous. Stochastic electrodynamics (SED), which is based on the existence of zero-point energy of value $\hbar\omega/2$, is another completely continuous theory, which has developed as a rival to the purely discrete theory of the quantum with energy $\hbar\omega$.

Again, the Klein–Gordon equation, based on potential energy, is effectively a 'discrete' one, in the space-time sense, whereas the Dirac equation, like the Schrödinger equation, based on kinetic energy, is effectively 'continuous'. 'Discrete' (or steady state) energy equations employ terms which are twice the size of those describing 'continuous' (or changing) conditions and the distinction is transferred into quantum mechanics with the quantum energy equations which are based on classical ones. The choice between the factors 1 and $\frac{1}{2}$ for spin, and other related quantities, seems to be made at the same time as that between time-like and space-like equations, and between discrete and continuous physics. In fact, the $\frac{1}{2}$ ratio between the spins of fermions and bosons provides a classic instance in which there are alternative explanations using *any one* of the three fundamental dualities.

Duality	Method
conserved/nonconserved	**potential energy**/kinetic energy
real/complex	**nonrelativistic**/relativistic
nondimensional/dimensional	**commutative**/anticommutative

There is, of course, only one system, whatever the description, and both options have to incorporate the alternative in some way. Though a single duality separates alternative theories, such as Heisenberg and Schrödinger, or quantum mechanics and stochastic electrodynamics, it is invariably open to more than one interpretation because each pair of parameters is always separated by two distinct dualities, and the separate interpretations ultimately act together when we consider a phenomenon in relation to its place in the overall 'environment' of the physical universe.

The Schrödinger approach is a continuous one based on $\frac{1}{2}\hbar$, but incorporates discreteness (based on $\hbar$) in the process of measurement — the 'collapse of the wavefunction'. The Heisenberg approach assumes a discrete system based on $\hbar$, but incorporates continuity (and $\frac{1}{2}\hbar$) in the process of measurement, via the uncertainty principle and zero-point energy. There is always a route by which $\frac{1}{2}\hbar\omega$ in one context can become $\hbar\omega$ in another. So, in black-body radiation, the spontaneous emission of energy of value $\hbar\omega$ combines the effects of $\frac{1}{2}\hbar\omega$ units of energy provided by both oscillators and zero-point field. A fermionic object on its own shows changing behaviour, requiring an integration which generates a factor $\frac{1}{2}$ in the kinetic energy term, and a sign change when it rotates through 2π, while a conservative 'system' of object plus environment shows unchanging behaviour, requiring a potential energy term which is twice the kinetic energy.

The $\frac{1}{2}\hbar\omega \rightarrow \hbar\omega$ transition for black body radiation can also be explained in terms of radiation reaction, which is connected again with the distinction between the relativistic and rest masses of an object. Rest mass effectively defines an isolated object with *kinetic* energy. *Relativistic* mass, on the other hand, already incorporates the effects of the environment. For a photon, which has no rest mass, and only a relativistic mass, the energy mc^2 behaves exactly like a classical potential energy term, as when a photon gas produces the radiation pressure $\rho c^2/3$. We take into account both action and reaction because the doubling of the value of the energy term comes from doubling the momentum when the photons rebound from the walls of the container, or, alternatively, are absorbed and re-emitted. Exactly, the same thing happens with radiation reactions, thus explaining an otherwise 'mysterious' doubling of energy from $\frac{1}{2}\,h\nu$ to $h\nu$. In a more classical context, Feynman and Wheeler find a doubling of the contribution of the retarded wave in electromagnetic theory, at the expense of the advanced wave, by assuming that the vacuum behaves as a perfect absorber and reradiator of radiation. This is effectively the same as privileging matter over vacuum, or one direction of time.

Radiation reactions are basically just another version of Newton's third law. For the anomalous magnetic moment of the electron, many of the same results are also explained by special relativity, but, as shown by C. K. Whitney,[35] the correct magnetic moment for the electron is obtained without relativity, by treating the transmission of light as a two-step process involving absorption and emission, which is, again, action and reaction. Two-step processes also remove those SR paradoxes which involve apparent reciprocity; in effect, SR, by including only one side of the calculation, effectively removes reciprocity, and so leads to such things as asymmetric ageing in the twin paradox.

The 'environment' can apply to either a material or vacuum contribution. The duality here connects with the boson/fermion distinction and the spin 1 or spin $\frac{1}{2}$ division, supersymmetry, vacuum polarization, pair production, renormalization, *zitterbewegung*, *etc.* The halving of energy in 'isolating' the fermion from its vacuum or material 'environment' is the same process as mathematically square-rooting the quantum operator via the Dirac equation. Energy principles determine that all fermions, in whatever circumstances, may be regarded either as isolated spin $\frac{1}{2}$ objects or as spin 1 objects in conjunction with some particular material or vacuum environment, or, indeed, the 'rest of the universe'; and fermions with spin $\frac{1}{2}$ automatically become spin 1 particles when taken in conjunction with their environment, whatever that may be.

In the Berry phase examples (Jahn–Teller effect, Aharonov–Bohm effect, *etc.*), the spin-$\frac{1}{2}$, $\frac{1}{2}$-wavelength-inducing nature of the fermionic state is a product of discreteness in both the fermion (and its charge) and the space in which it acts. The very act of creating a discrete particle requires a splitting of the continuum vacuum into *two* discrete halves, as with a rectangle into two triangles or (relating the concept of discreteness to that of dimensionality) two square roots of 0. That the doubling mechanism also applies in purely mathematical, as well as in physical, contexts is evident from the topological explanation of the Aharonov–Bohm effect, though the physical and mathematical applications must ultimately have the same origin.

The very concept of duality also implies that the actual processes of counting and generating numbers are created at the same time as the concepts of discreteness, nonconservation, and orderability are separated from those of continuity, conservation, and nonorderability. The mathematical processes of addition and squaring are, in effect, 'created' at the same time as the physical quantities to which they apply, while all the other fundamental mathematical concepts and processes (e.g. the Dedekind cut) are, in some way, defined by dualling. The factor 2 thus expresses dualities which are fundamental to the creation of both mathematics and physics, and duality provides a philosophy on which both can be based.

Square-rooting and halving in mathematics have an intimate relationship, which is manifested physically in the relation between vector spin terms of bosons and fermions and their respective uses of double or single nilpotent operators, in addition to the halving approximation used to find the kinetic energy term in the binomial expansion for relativistic mass. This relationship is determined entirely by the fact that 3D Pythagorean addition is a dualistic process with a numerical doubling arising from noncommutativity, and this applies to both the vector operators used for space and momentum, and the quaternion operators used in the Dirac nilpotent.

A few cases of the general idea of duality are of particular interest. At 0.25, the idealised electroweak mixing parameter $\sin^2\theta_W$ calculated from $Tr\ (t_3^2)/Tr(Q^2)$ becomes equivalent to making the weak charge value twice that of the electric charge at the energy at which the mixing takes place. The ultimate reason for this is the fact that the quantum number for weak charge (t_3) is half that for electric charge (Q), because of the $SU(2)$ nature or dipolarity introduced by the complexifying factor i, while the compensating factors of weak isospin and single-handedness contrive to halve the respective numbers of electric and weak states simultaneously — which is, of course, no coincidence, because both are aspects of weak dipolarity. Ultimately, then, the weak charge has ideally twice the magnitude of the electric charge at the energy at which they are mixed because the weak charge is dipolar, and the weak charge is dipolar because it is complexified. Could the magnitude of the *strong* charge, at the same energy of interaction, be *twice that of the weak charge* through a further doubling effect due to dimensionalization? It is certainly of this order at the energy of the electroweak scale. Among other less obvious examples of the factor 2 dual process are the magnetic flux quantum term $\hbar/2e$ in the time-dependent voltage of the

Josephson effect, $U(t) = (\hbar/2e)\partial\phi/\partial t$, produced by a *changing* phase difference, ϕ, and the pair production producing only one observable fermion in Hawking radiation.

As entertainment, we can look at the famous equation $e^{i\pi} = -1$ and see that it is a remarkable case of a combination of all three dualities! If we take mass as being conserved, real and nondimensional, and with positive real unit 1, we will see that e is defined by differentiation (nonconserved), i is imaginary, and π is defined by 3-dimensionality, and that these 3 act together to produce a collective dual to positive unit 1. (The combination has parallels to CPT in physics, which, of course, combines the properties of the 3 quantities which are dual to mass: charge, space and time.)

In terms of the mathematical structure generated by the universal rewrite structure, it would be possible to classify the physical phenomena involving the factor 2 as resulting from the three distinct mathematical processes, which can be identified as *conjugation*, *complexification* and *dimensionalization*, and which are manifested, respectively, through opposite signs (or equivalent), the distinction between real and imaginary components, and the introduction of cyclic dimensionality.

Physical phenomenon	Mathematical process which causes it
action and reaction	conjugation
commutation relations	dimensionalization
absorption and emission	conjugation
object and environment	conjugation
relativity	complexification
electroweak mixing	complexification
the virial relation	conjugation
continuity and discontinuity	conjugation/dimensionalization

It may be that, in larger scale systems, which particular duality is invoked depends on the point at which the first zero-totality alphabet is defined.

Overall, the factor 2 appears as either the link between the continuous and discrete physical domains, or between the changing and the fixed, or the real and imaginary (orderable and nonorderable), the three dualities of the Klein-4 parameter group, and, in every physical instance, between more than one of these. Duality seems to be the necessary result of any attempt to create singularity. The infinite imaging of the fermionic state in the vacuum and the nonlocal connection between all the state vectors in the entire universe, or infinite entanglement of all nilpotent fermionic states, described mathematically in terms of Hilbert space, provide an extension of the dualling process to infinity. Ultimately, it would seem, duality is not merely a 'component' of physics but an expression of the fundamental nature of physics itself.

10.6 Anticommutativity and the factor 3

3-dimensionality, one of the most profound and fundamental concepts in physics, has its origin in ideas of anticommutativity, which may be antecedent to the concept of numbers. It seems to be responsible for all discreteness in physical systems, and in particular for quantization, as well as for symmetry breaking between the forces, for many significant aspects of particle structure, and for most of the manifestations of the number 3 that are considered fundamental in physics. The Dirac equation is specially structured to accommodate it. We have 3 dimensions of space, 3 nongravitational interactions, 3 fundamental symmetries (C, P and T), 3 conserved dynamical quantities (momentum, angular momentum and energy), 3 quarks in a baryon, 3 generations of fermions (which can be related to C, P and T).

No other dimensionality, not even that of '4-dimensional' space-time, has any fundamental physical significance. The connection between space and time is basically 3-dimensional ($\boldsymbol{k}$, $\boldsymbol{i}$, $\boldsymbol{j}$), and not privileged with respect to mass and charge. Time is *not* part of space, but of *another* 3-dimensionality, though the differences can be masked when we take the scalar product. An ordinary connection between space and time, not mediated by this second 3-dimensionality, leads to wave–particle duality, where one parameter has to take in the other's physical aspects. Even within the higher dimensionalities of the Dirac algebra, the nilpotent structure shows its fundamental 3-dimensionality, and it is this inherent 3-dimensionality which allows us to develop a fully renormalizable formal theory of quantum gravitational inertia.

Just as there are two 3-dimensional structures that are fundamental in nature, so there are also two manifestations of 3-dimensionality, the nonconserved and the conserved. The nonconserved has an unbroken symmetry with axes not separately identifiable; in the conserved case, the symmetry is broken or chiral, with axes separately identifiable. The symmetry-breaking always has the same structure: one term is complexified; one term is associated with an unbroken 3-dimensionality; and the remaining term is purely scalar. The whole structure is invariably nilpotent, reflecting the conserved nature, and either the scalar or the complexified (pseudoscalar) term becomes redundant, except as a 'book-keeper'.

Our quantized, i.e. 3-dimensional, picture denies us the opportunity of representing time as a fourth dimension or giving it status as a physical observable. In a *3-dimensional* theory, time occupies the place of the 'book-keeper', as energy does in the Dirac state, the quantity which preserves conservation or conjugation, but adds only the information of $+$ or $-$. We only know the direction of the sequence that preserves causality, not a *measure* of time in the same sense as we measure space, in the same way as energy only tells us whether the system is a fermion or antifermion. This fact is well known as a stumbling block to proponents of a quantum theory of gravity, which automatically incorporates time as a physical fourth dimension. The fact that the number system we use in mathematics may have a 3-dimensional origin is of profound significance. It means that we can't arbitrarily choose the number of dimensions we apply to quantities like space and time without contradicting the principles on which these concepts, and related ones, such as quantization and conservation, are founded.

The separate roles for the three axes in a 3-dimensional system with identifiable components has a remarkable similarity with the processes involved in creating the infinite algebra of the rewrite system. The role of $\boldsymbol{j}$ is essentially that of complexification, the beginning of a new and as of yet incomplete quaternion system. The role of $\boldsymbol{i}$ is to introduce dimensionalization, while $\boldsymbol{k}$ is restricted to the 'book-keeping' role of conjugation or conservation. These also run parallel to the roles of scalar, vector and pseudoscalar quantities, which an extra i factor has transformed from the sequence pseudoscalar, quaternion, scalar. It is this parallelism which makes it possible to create a closed parameter system with zero totality and in-built repetition.

pseudoscalar	quaternion	scalar
scalar	vector	pseudoscalar
mass	space	time
m	$\mathbf{p}$	E
τ	$\mathbf{r}$	t
e	s	w
C	P	T

j	i	k
magnitude	direction	orientation
complexification	dimensionalization	conjugation
complexification	dimensionalization	conservation

The 'dimensional' term here is in the second column and the 'book-keeping' term in the third. The first two rows are related to each other by the multiplication of i, as mentioned before. It may be that we can also include momentum–angular momentum–energy and space translation–space rotation–time translation. The last row refers to the properties of the parameter group.

10.7 Symmetry and self-organization

Physics uses many groups, geometries, algebras, and displays many symmetries, but the only pure symmetries that matter are all based on 2 and 3, and the most fundamental broken symmetries are based on 5. The table of numbers related to particles in Chapter 9 includes practically all the integers that are fundamentally important in physics and in many other scientific areas, including biology, and they are all based on 2, 3 and 5, and these, in turn, all emerge from the universal rewrite system. It may be that this system is a general description of process which applies to mathematics, computation, physics, chemistry, biology, and any self-organizing system. Investigations in a considerable number of areas appear to show signs of the same structures and symmetries, in exactly the way we said that nature made it possible for us to understand things way outside our experience.

The nilpotent relation between the defined system and the rest of the universe, which emerges in the nilpotent form of quantum mechanics could possibly be an indicator of a process of self-organization which occurs at increasing levels of complexity. The particular characteristics of this type of self-organizing include double 3-dimensionality; a 5-fold broken symmetry; geometric phase; uniqueness of the objects and unique birthordering; irreversibility; dissipation; chirality, a harmonic oscillator mechanism, *zitterbewegung*; fractality of dimension 2; the holographic principle and quantum holography. Research is ongoing in this area, with special (but not unique) reference to biology where it is clear that some sort of information process is a bigger driver than the specific chemical processes involved.

We have seen in our investigations of the foundations of physical law that the separate concept of the 'physical', meaning the 'tangible' or 'material', has no fundamental validlity. Only the purely abstract has any real claim to be a true description of nature. This could have been suspected as soon as quantum mechanics required us to use such concepts as vacuum and nonlocality at the same time as localised material particles. No amount of explanation can transform such abstract notions into 'tangible' realities, and yet we cannot avoid using them. Even 'real particles', however, are nothing but mathematical points, and it is here that we see that physics at the fundamental level effectively assumes the character of a pure mathematics, though of a very special nature.

Physical laws are founded on the intrinsically *mathematical* definition of two algebraic or geometric vector spaces. In effect, translational and rotational symmetries ensure that a point has no meaning as a component of a single space. Combining two spaces to ensure zero totality, however, forces us to break the symmetry of one of the two spaces, imposing a *conservation* condition on the space with the broken symmetry. Algebraically, this symmetry-breaking has the exact character of the $SU(3) \times SU(2) \times U(1)$ structure of the Standard Model. Purely generic laws like Maxwell's equations and the laws for strong and weak interactions become the working out of the conditions necessary to produce this algebra. There is no separate 'physical' content. Nilpotent quantum mechanics defines the conditions for the point to exist at all, with the point acting simultaneously as 'source' and 'sink' with respect to the rest of the universe, or the 'vacuum' which forms its mirror image. Fundamental particles seem structured to define the exact conditions under which such points can exist, and the purely abstract Hilbert space in which the points are created in an indefinite sequence seems to be the natural endproduct of the universal rewrite system. If nature, and specifically physics, can be seen effectively as the working out of a branch of pure mathematics, then we may well ask what branch of mathematics it constructs, and the answer would seem to be the mathematics of *uniqueness*, or a sequence which forms a unique birthordering based on endless zero totalities. Because it is unique we can never create an equivalent representation of the whole, but we can use the nilpotent construction to show that it *is* unique and we can find partial representations in terms of repetitive sequences which apply to our *local* observations. Unique does not mean random, and the local order enables us to use pattern recognition to show that a birthordering, however unknowable in absolute terms, has a structure that we can recognise.

The purpose of this book has been to show that the foundations of physics is a very different subject from anything we might previously have imagined. It requires its own methodology and philosophy, and even to a certain extent, its own mathematics. It requires, in addition, a great deal of inductive thinking. Yet it is also rigorous and leads to mathematically and experimentally testable results. Besides creating its own results, it also adds significantly to physics worked out at a greater level of complexity. Its methods and symmetries can suggest answers where none are available by any other means, and resolve anomalies where this had previously seemed impossible. I would like to think that anyone who has followed the argument presented here will see that the possibilities opened up are almost limitless with many opportunities for significant research.

References

1. R. Feynman, W. Morinigo, and W. Wagner, *Feynman Chapters on Gravitation* (for academic year 1962–63), Addison-Wesley Publishing Company, 1995, p. 10.
2. P. Atkins, *Creation Revisited*, Harmondsworth, 1994, p. 23.
3. D. Hestenes, *Space-Time Algebras*, Gordon and Breach, 1966. See also W. Gough, Mixing scalars and vectors — an elegant view of physics, *Eur. J. Phys.* **11**, 326–3, 1990.
4. A. Robinson, *Non-Standard Analysis*, Princeton University Press, 1996, original publication, 1966.
5. O. Lodge, On the identity of energy: In connection with Mr. Poynting's paper on the transfer of energy in an electromagnetic field, and on the two fundamental forms of energy, *Phil. Mag.* 5th series, **19**, 482–7, 1885.
6. G. J. Whitrow, *The Natural Philosophy of Time*, Nelson, London, 1961, pp. 135–57. P. Coveney and R. Highfield, *The Arrow of Time*, London, 1990, pp. 28, 143–4, 157.
7. L. H. Kauffman, Non-commutative worlds, *New Journal of Physics* **6**, 2–46, 2004.
8. P. A. M. Dirac, *The Principles of Quantum Mechanics*, fourth edition, Clarendon Press, Oxford, 1958.
9. E. Schrödinger, Über die kräftefreie Bewegung in der relativistischen Quanten-mechanik, *Sitz. Preuss. Akad. Wiss. Phys.-Math. Kl.* **24**, 418–28, 1930.
10. P. Rowlands, *Zero to Infinity: The Foundations of Physics*, World Scientific, 2007, Chapters 6 and 11.
11. P. West, *Introduction to Supersymmetry and Supergravity*, World Scientific, 1986, p. 15.
12. J. C. Baez and J. Huerta, The strangest numbers in string theory, *Scientific American*, May 2011, 18.
13. P. Rowlands, *Zero to Infinity: The Foundations of Physics*, World Scientific, 2007, Chapter 18.
14. A. Faroggi, Lectures, University of Liverpool.

15. S. Lloyd, Computational capacity of the universe. *Phys. Rev. Lett.* **88** (23), 237901, 2002. V. Giovannetti, S. Lloyd and L. Maccone, Quantum-enhanced measurements: Beating the standard quantum limit, *Science* **306**, 1330–6, 2004.
16. C. R. Moon, L. S. Mattos, B. K. Foster, G. Zeltzer and H. C. Manoharan, *Nature Nanotechnology* **4**, 167–72, 2009.
17. I. Newton, *Opticks*, 1704, Query 1.
18. E. Cartan, *Ann. Ecole Norm.* **40**, 325, 1923; **41**, 1, 1924.
19. C. M. Will, Henry Cavendish, Johann von Soldner, and the deflection of light, *American Journal of Physics* **56**, 413–5, 1988.
20. J. Antoniadis *et al.*, A massive pulsar in a compact relativistic binary, *Science* **340**(6131), 499, 26 April 2013.
21. P. Rowlands, *Zero to Infinity: The Foundations of Physics*, World Scientific, 2007, p. 481.
22. H. Kolbentsvedt, *Amer. J. Phys.* **56**, 523–4, 1988.
23. D. W. Sciama, *Monthly Notices RAS*, **113**, 34, 1953; *The Physical Foundations of General Relativity*, Heinemann Educational Publisher, 1972, pp. 36 ff.
24. P. Rowlands, *A Revolution Too Far: The Establishment of General Relativity*, PD Publications, 1994.
25. Planck Collaboration I, arxiv:1303.5062; XVI, arxiv: 1303.5076.
26. M. de Souza and R. N. Silveira, Discrete and finite general relativity, gr-qc/980140, 1998.
27. M. Y. Han and Y. Nambu. Three-triplet model with double $SU(3)$ symmetry. *Phys. Rev.* **139B**, 1006–10, 1965.
28. R. B. Laughlin, Anomalous quantum Hall effect: An incompressible field with fractionally charged excitations, *Phys. Rev. Lett.* **50**, 1395–8, 1983.
29. H. Georgi and S. L. Glashow, Unity of all elementary-particle forces, *Phys. Rev. Lett.* **32**, 438–41, 1974.
30. S. Weinberg, *The Quantum Theory of Fields*, 2 vols., Cambridge University Press, 1996, Vol. II, pp. 327–32.
31. A. Garrett Lisi, An exceptionally simple theory of everything, arXiv:hep/0711.0770v1.
32. V. Hill and P. Rowlands, The numbers of nature's code, *International Journal of Computing Anticipatory Systems* **25**, 160–175, 2010.
33. P. Marcer and P. Rowlands, Information, Bifurcation and Entropy in the Universal Rewrite System, Liège, August 2011, *International Journal of Computing Anticipatory Systems* (in press).
34. R. Landauer, Computation and physics: Wheeler's meaning circuit?, *Foundations of Physics* **16**, 551–564, 1986.
35. C. K. Whitney, How can paradox happen?, *Proceedings of Conference on Physical Interpretations of Relativity Theory VII*, British Society for Philosophy of Science, London, September 2000, pp. 338–51.

Index

www.ingramcontent.com/pod-product-compliance
Lightning Source LLC
Chambersburg PA
CBHW050551190326
41458CB00007B/2004
* 9 7 8 9 8 1 4 6 1 8 3 7 3 *